THE SOLAR SYSTEM

By

Steve Parker

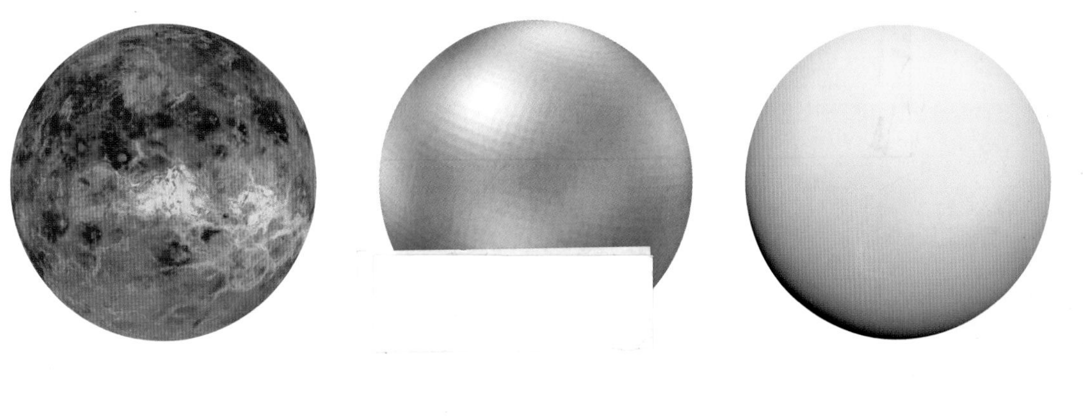

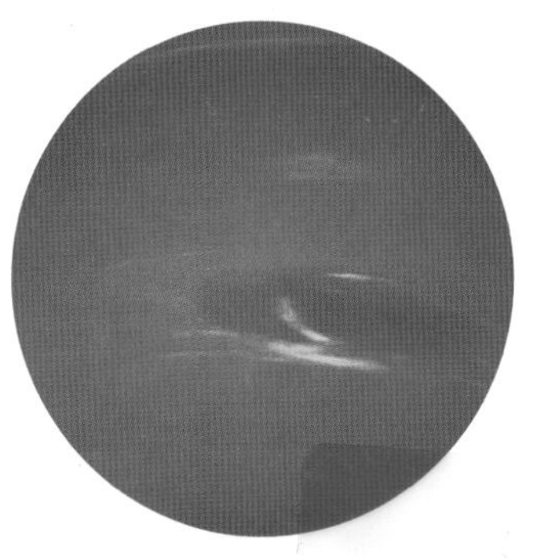

First published in Great Britain in 2006 by **ticktock Media Ltd.**,
Unit 2, Orchard Business Centre, North Farm Road, Tunbridge Wells, Kent, TN2 3XF
ISBN 1 86007 857 5 pbk
Printed in China
A CIP catalogue record for this book is available from the British Library.

Picture Credits: All images NASA except: Alamy: 7t, 19b, 20r, 33c, 38l, 39tr, 48b, 50t, 52r. Bridgeman Art Library: 2 & 9t, 8-9 all, 10c, b &br, 11 all, 27t, 31c, 35t, 40c. Kobal Collection: 23b.

CONTENTS

HOW TO USE THIS BOOK

JUST THE FACTS, THE SOLAR SYSTEM is an easy-to-use, quick way to look up facts about our Solar System. Every page is packed with cut-away diagrams, charts, scientific terms and key pieces of information. For fast access to *just the facts*, follow the tips on these pages.

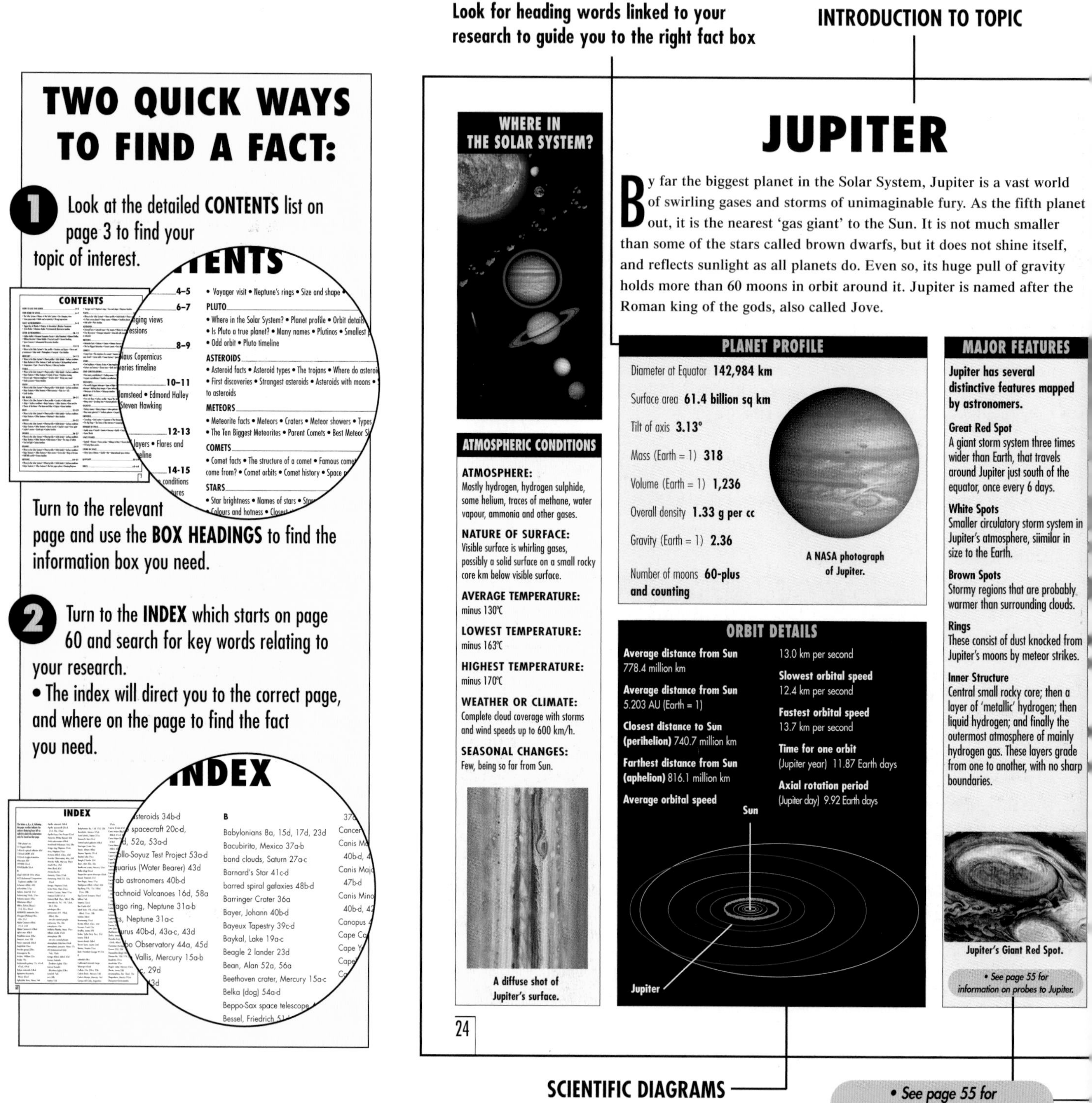

BOX HEADINGS
Look for heading words linked to your research to guide you to the right fact box

INTRODUCTION TO TOPIC

TWO QUICK WAYS TO FIND A FACT:

1 Look at the detailed **CONTENTS** list on page 3 to find your topic of interest.

Turn to the relevant page and use the **BOX HEADINGS** to find the information box you need.

2 Turn to the **INDEX** which starts on page 60 and search for key words relating to your research.

• The index will direct you to the correct page, and where on the page to find the fact you need.

JUPITER

By far the biggest planet in the Solar System, Jupiter is a vast world of swirling gases and storms of unimaginable fury. As the fifth planet out, it is the nearest 'gas giant' to the Sun. It is not much smaller than some of the stars called brown dwarfs, but it does not shine itself, and reflects sunlight as all planets do. Even so, its huge pull of gravity holds more than 60 moons in orbit around it. Jupiter is named after the Roman king of the gods, also called Jove.

WHERE IN THE SOLAR SYSTEM?

ATMOSPHERIC CONDITIONS

ATMOSPHERE: Mostly hydrogen, hydrogen sulphide, some helium, traces of methane, water vapour, ammonia and other gases.

NATURE OF SURFACE: Visible surface is whirling gases, possibly a solid surface on a small rocky core km below visible surface.

AVERAGE TEMPERATURE: minus 130°C

LOWEST TEMPERATURE: minus 163°C

HIGHEST TEMPERATURE: minus 170°C

WEATHER OR CLIMATE: Complete cloud coverage with storms and wind speeds up to 600 km/h.

SEASONAL CHANGES: Few, being so far from Sun.

A diffuse shot of Jupiter's surface.

24

PLANET PROFILE

Diameter at Equator **142,984 km**

Surface area **61.4 billion sq km**

Tilt of axis **3.13°**

Mass (Earth = 1) **318**

Volume (Earth = 1) **1,236**

Overall density **1.33 g per cc**

Gravity (Earth = 1) **2.36**

Number of moons **60-plus and counting**

A NASA photograph of Jupiter.

ORBIT DETAILS

Average distance from Sun 778.4 million km

Average distance from Sun 5.203 AU (Earth = 1)

Closest distance to Sun (perihelion) 740.7 million km

Farthest distance from Sun (aphelion) 816.1 million km

Average orbital speed 13.0 km per second

Slowest orbital speed 12.4 km per second

Fastest orbital speed 13.7 km per second

Time for one orbit (Jupiter year) 11.87 Earth days

Axial rotation period (Jupiter day) 9.92 Earth days

Sun

Jupiter

MAJOR FEATURES

Jupiter has several distinctive features mapped by astronomers.

Great Red Spot
A giant storm system three times wider than Earth, that travels around Jupiter just south of the equator, once every 6 days.

White Spots
Smaller circulatory storm system in Jupiter's atmosphere, siimilar in size to the Earth.

Brown Spots
Stormy regions that are probably warmer than surrounding clouds.

Rings
These consist of dust knocked from Jupiter's moons by meteor strikes.

Inner Structure
Central small rocky core; then a layer of 'metallic' hydrogen; then liquid hydrogen; and finally the outermost atmosphere of mainly hydrogen gas. These layers grade from one to another, with no sharp boundaries.

Jupiter's Giant Red Spot.

• See page 55 for information on probes to Jupiter.

SCIENTIFIC DIAGRAMS
Clear, accurate diagrams explain difficult astronomic concepts.

• See page 55 for information on probes to Jupiter.

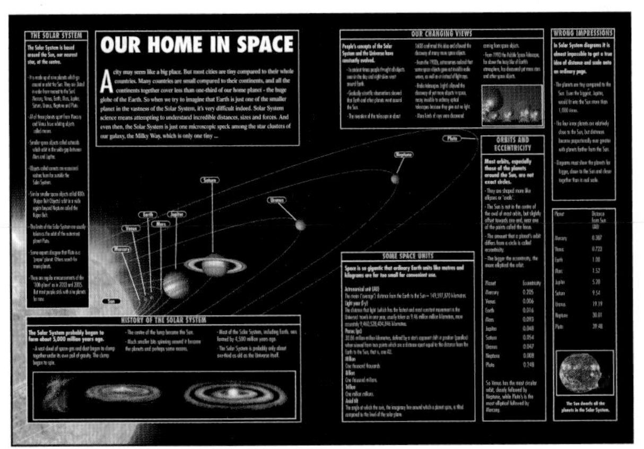

6-7 Our Home in Space

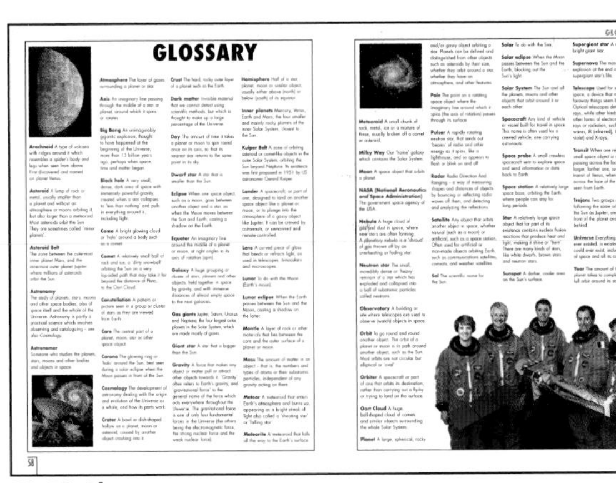

58-59 Glossary

OTHER FEATURES

- **BELTS** Strips of dark clouds which snake along from west to east (left to right) and change through the years.
- **ZONES** Lengths of light-coloured clouds which change like the darker belts. Zones contain higher clouds than belts. Blue-tinted clouds are lowest and warmest.
- **TURBULENCE** Belts sometimes move in the opposite direction to their neighbouring zones, creating swirling patterns of storms and turbulence along their edges.

MOON RECORDS

- Ganymede is the largest moon in the Solar System.
- Callisto is the most heavily cratered object.
- Io probably has the most volcanic activity.

A NASA collage showing IO's red and black coloured lava flows and yellow sulphur patches.

JUPITER'S RINGS

Main Ring
Gossamer Rings
Halo
Amalthea
Adrastea
Metis
Thebe

- **HALO RING** Faint inner ring 10,000 km thick.
- **MAIN RING** Dust from Adrastea and Metis, 100 km thick.
- **INNER GOSSAMER RING** Dust from Amalthea, 2,600 km thick.
- **OUTER GOSSAMER RING** Dust from Thebe 8,800 km thick
- **FAINT OUTER RING** Rings beyond outer Gossamer Ring, 8,800 km thick

TRUE GIANT

Jupiter has more than twice as much mass ('weight') than all the other eight planets added together. However it would probably need to be 50 times heavier to start burning like a true star.

SPEED SPIN

Jupiter is not only the largest planet, it also spins around the fastest, once in less than 10 Earth hours. The spinning speed of the upper atmosphere at the equator is 5 minutes faster than at the poles, so the atmosphere is continually being twisted and torn.

JUPITER'S MOONS

On 7-11 January 1610 Galileo discovered Jupiter's four main moons, now known as Galilean moons, following their orbits across the face of the planet. This was direct evidence that the Earth was not at the centre of everything, and strengthened his idea that planets like Earth and Jupiter probably revolved around the Sun.

MOON (or group)	DIAMETER	DISTANCE FROM JUPITER
Inner group (Metis, Adrastea, Amalthea, Thebe)	Four small moons less than 200 km across	Less than 200,000 km
Io	3,642 km	421,700 km
Europa	3,122 km	671,030 km
Ganymede	5,262 km	1,070,000 km
Callisto	4,820 km	1,882,000 km
Themisto	8 km	7,392,000 km
Himalia group	Most under 100 km	11-12 million km
Ananke group	Most under 30 km	21 million km
Carme group	Most under 5 km	23-24 million km
Pasiphaë	Small outermost moons	23 million km

JUPITER TIMELINE

3,000 years ago
Jupiter was known to the Greeks and then the Romans.

1,500 years ago
In Ancient China Jupiter was known as the 'Wood Star'.

1610
Galileo observed Jupiter's four largest moons.

1665
The Great Red Spot was first observed.

1690
Giovanni Domenico Cassini noticed that the upper atmosphere takes longer to spin at the poles than around the equator.

1973
US Pioneer 10 probe flew past Jupiter.

1979
Voyager 1 flew past taking spectacular photographs. Voyager 2 achieved similar results.

1992
Ulysses probe passes by Jupiter on its way to the Sun taking measurements.

1994
Parts of comet Shoemaker-Levy 9 hit Jupiter in July, photographed by the approaching Galileo space probe.

1995
Galileo became the first probe to go into orbit around Jupiter, starting on 7 December. On the same day an atmosphere probe it had already released parachuted 150 km into the atmosphere, collecting information.

1996-2003
Galileo continued its studies of Jupiter and its nearer moons.

2000
The Great Red Spot had shrunk to about half the size it measured in 1900.

2000
Cassini probe passes on way to Saturn.

2003
Galileo finally plunged into the clouds in September.

2007
The New Horizons probe is due to fly past, on its way to Pluto.

2010
The US Jupiter probe Juno is scheduled for launch. Juno will orbit over Jupiter's poles.

25

TIMELINES

Important events are listed in chronological order.

For fast access to facts in the timelines, look for key words in the headings.

1992
Ulysses probe passes by Jupiter on its way to the Sun taking measurements.

GLOSSARY

- **A GLOSSARY of words and terms used in this book begins on page 58.**
- **The glossary words provide additional information to supplement the facts on the main pages.**

JUST THE FACTS

Each topic box presents the facts you need in short, easy to follow information.

LINKS

Look for the purple links throughout the book. Each link gives details of other pages where related or additional facts can be found.

OUR HOME IN SPACE

A city may seem like a big place. But most cities are tiny compared to their whole countries. Many countries are small compared to their continents, and all the continents together cover less than one-third of our home planet - the huge globe of the Earth. So when we try to imagine that Earth is just one of the smaller planet in the vastness of the Solar System, it's very difficult indeed. Solar System science means attempting to understand incredible distances, sizes and forces. And even then, the Solar System is just one microscopic speck among the star clusters of our galaxy, the Milky Way, which is only one tiny ...

THE SOLAR SYSTEM

The Solar System is based around the Sun, our nearest star, at the centre.

- It is made up of nine planets which go around or orbit the Sun. They are (listed in order from nearest to the Sun) Mercury, Venus, Earth, Mars, Jupiter, Saturn, Uranus, Neptune and Pluto.
- All of these planets apart from Mercury and Venus have orbiting objects called moons.
- Smaller space objects called asteroids which orbit in the wide gap between Mars and Jupiter.
- Objects called comets are occasional visitors from far outside the Solar System.
- Similar smaller space objects called KBOs (Kuiper Belt Objects) orbit in a wide region beyond Neptune called the Kuiper Belt.
- The limits of the Solar System are usually taken as the orbit of the outermost planet Pluto.
- Some experts disagree that Pluto is a 'proper' planet. Others search for more planets.
- There are regular announcements of the '10th planet' as in 2003 and 2005. But most people stick with nine planets for now.

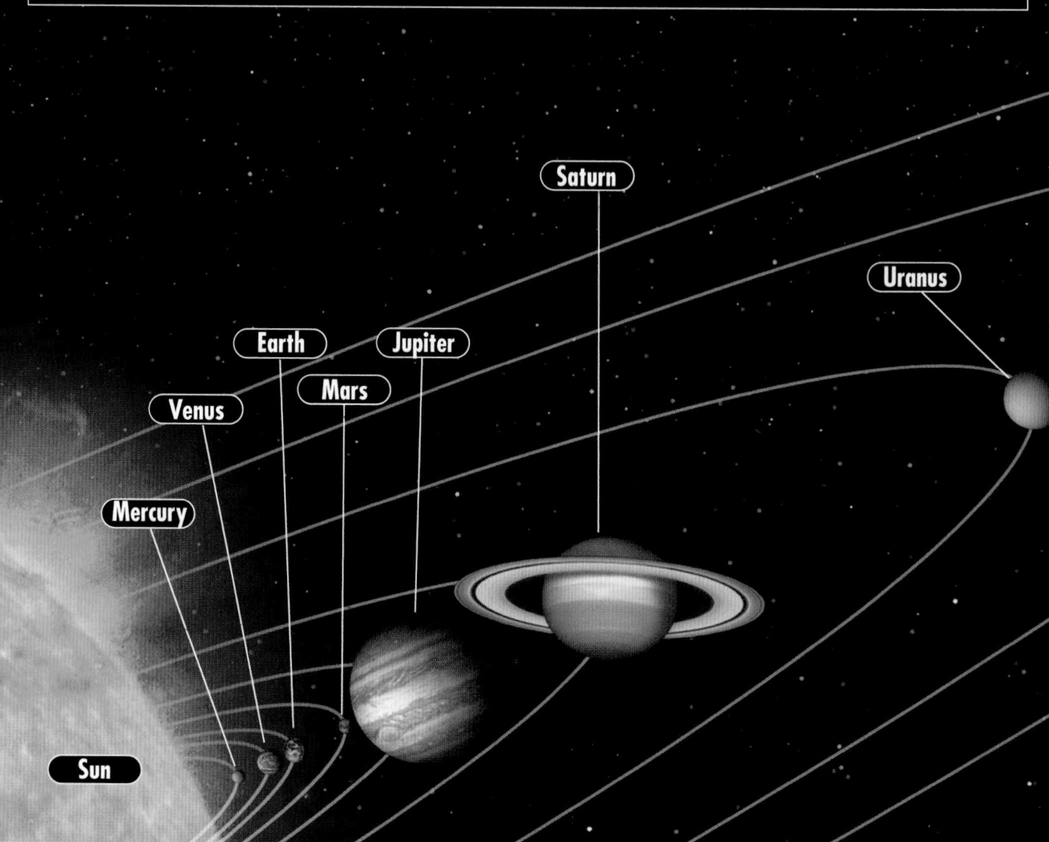

HISTORY OF THE SOLAR SYSTEM

The Solar System probably began to form about 5,000 million years ago.

- A vast cloud of space gas and dust began to clump together under its own pull of gravity. The clump began to spin.
- The centre of the lump became the Sun.
- Much smaller bits spinning around it became the planets and perhaps some moons.
- Most of the Solar System, including Earth, was formed by 4,500 million years ago.
- The Solar System is probably only about one-third as old as the Universe itself.

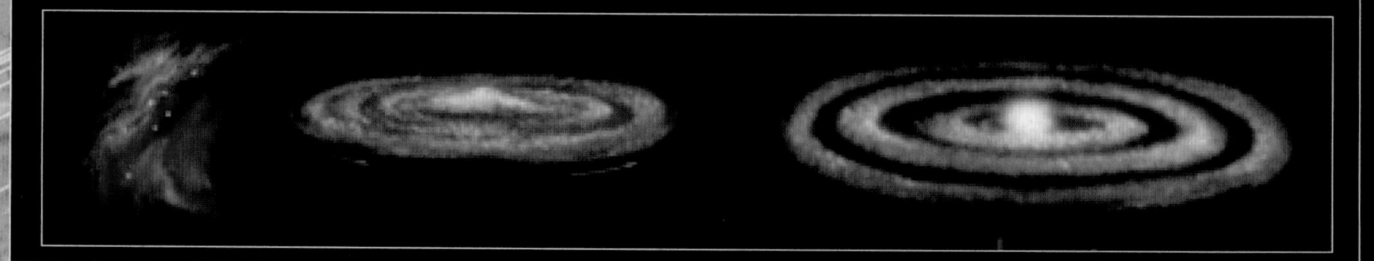

OUR CHANGING VIEWS

People's concepts of the Solar System and the Universe have constantly evolved.

- In ancient times people thought all objects seen in the day and night skies went around Earth.
- Gradually scientific observations showed that Earth and other planets went around the Sun.
- The invention of the telescope in about 1608 confirmed this idea and allowed the discovery of many more space objects.
- From the 1930s, astronomers realized that some space objects gave out invisible radio waves, as well as or instead of light rays.
- Radio telescopes (right) allowed the discovery of yet more objects in space, many invisible to ordinary optical telescopes because they give out no light.
- More kinds of rays were discovered coming from space objects.
- From 1990 the Hubble Space Telescope, far above the hazy blur of Earth's atmosphere, has discovered yet more stars and other space objects.

WRONG IMPRESSIONS

In Solar System diagrams it is almost impossible to get a true idea of distance and scale onto an ordinary page.

- The planets are tiny compared to the Sun. Even the biggest, Jupiter, would fit into the Sun more than 1,000 times.
- The four inner planets are relatively close to the Sun, but distances become proportionally ever greater with planets farther from the Sun.
- Diagrams must show the planets far bigger, closer to the Sun and closer together than in real scale.

Planet	Distance from Sun (AU)
Mercury	0.387
Venus	0.723
Earth	1.00
Mars	1.52
Jupiter	5.20
Saturn	9.54
Uranus	19.19
Neptune	30.01
Pluto	39.48

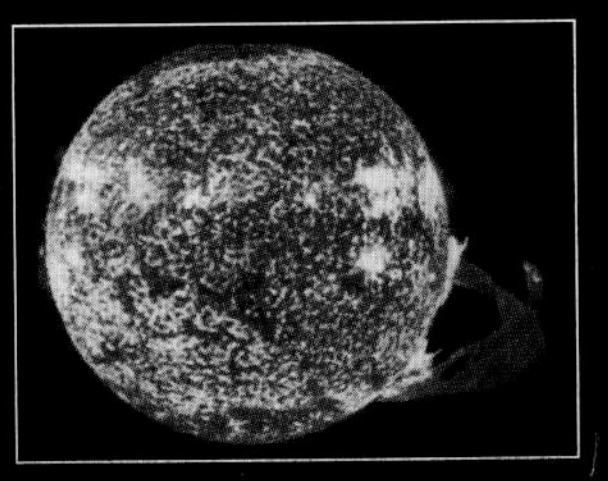

The Sun dwarfs all the planets in the Solar System.

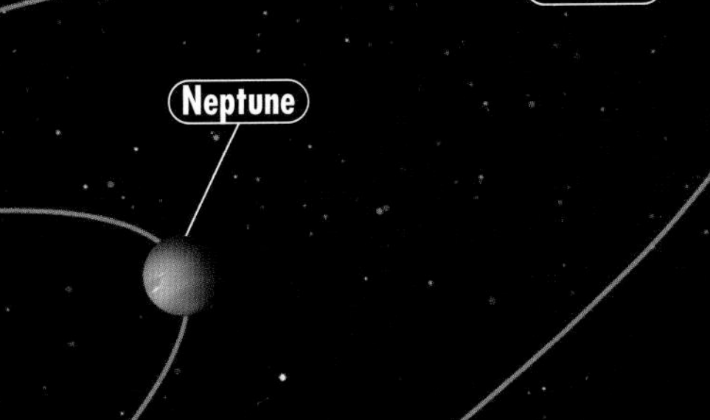

ORBITS AND ECCENTRICITY

Most orbits, especially those of the planets around the Sun, are not exact circles.

- They are shaped more like ellipses or 'ovals'.
- The Sun is not in the centre of the oval of most orbits, but slightly offset towards one end, near one of the points called the focus.
- The amount that a planet's orbit differs from a circle is called eccentricity.
- The bigger the eccentricity, the more elliptical the orbit.

Planet	Eccentricity
Mercury	0.205
Venus	0.006
Earth	0.016
Mars	0.093
Jupiter	0.048
Saturn	0.054
Uranus	0.047
Neptune	0.008
Pluto	0.248

So Venus has the most circular orbit, closely followed by Neptune, while Pluto's is the most elliptical followed by Mercury.

SOME SPACE UNITS

Space is so gigantic that ordinary Earth units like metres and kilograms are far too small for convenient use.

Astronomical unit (AU)
The mean ('average') distance from the Earth to the Sun – 149,597,870 kilometres.

Light year (l-y)
The distance that light (which has the fastest and most constant movement in the Universe) travels in one year, usually taken as 9.46 million million kilometres, more accurately 9,460,528,404,846 kilometres.

Parsec (pc)
30.86 million million kilometres, defined by a star's apparent shift in position (parallax) when viewed from two points which are a distance apart equal to the distance from the Earth to the Sun, that is, one AU.

Million
One thousand thousands.

Billion
One thousand millions.

Trillion
One million millions.

Axial tilt
The angle at which the axis, the imaginary line around which a planet spins, is tilted compared to the level of the solar plane.

EARLY ASTRONOMERS

The earliest astronomers were not interested in how the Universe worked. They needed to know when to plant or harvest crops, and when rivers would flood, and they used the movements of bodies in the heavens to make calendars, and to predict events in the future. Consequently they became astrologers, as well as astronomers. It was the ancient Greeks who first started to ask questions about the universe and how it worked. Their work was built on by the studies of great European astronomers from the 15th century onwards.

ASTRONOMICAL DISCOVERIES TIMELINE

27,000 years ago
Stone-age people made carvings and pictures of the Sun and Moon.

5,000 years ago

Sumerians record sightings of Mercury, which they call Ubu-idim-gud-ud.

Egyptians introduce a year of 365 days - the predecessor of our modern calendar.

4,000 years ago

Chinese astronomers make first record of Solar eclipse

4,000 years ago

Babylonian priests made some of the first records of astronomical observations.

3,500 years ago
Venus known to Babylonians.

2,455 years ago

Anaxagoras of Ancient Greece suggested the Sun is made of hot rocks.

2,360 years ago
Chinese astronomers may have spotted the moons of Jupiter.

2,265 years ago
Aristarchus proposed the Sun was the centre of the Solar System.

2,000 years ago
Jupiter and Saturn known to Greeks and Romans.

1,855 years ago
Ptolemy's Earth-centred view of the Solar System began to dominate for 1,500 years.

1543
Copernicus revived the suggestion of Aristarcus.

1608
Lippershey invented the optical telescope.

1609
Galileo began his space studies.

Kepler published his first and second laws of planetary motion.

HIPPARCHUS OF RHODES

Lifetime: Between 190 -120 BC.

Nationality: Greek.

Major Achievements:

- Hipparchus was believed to have catalogued over 800 stars, and also studied the motions of the Moon.
- He invented a brightness scale, subsequently developed by later generations of astronomers into a scale referred to as magnitude.
- Hipparchus calculated the length of the Earth's year to within 6 1/2 minutes.

An illustration of the early astonomer Hipparchus.

• See page 40 for more information on magnitude.

PTOLEMY OF ALEXANDRIA

Lifetime: 87-150 AD

Nationality: Greek.

Major Achievements:

- Ptolemy wrote many books containing Greek ideas and observations collected over the past 500 years, including *Almagest* ('The Greatest'/'The Mathematical Collection').
- Ptolemy described more than 1,000 stars in his books, including 48 different constellations.
- The astronomer also made early calculations of the size and distance of the Sun and Moon.
- Ptolemy devised a geocentric system with Earth at the centre of the Sun, Moon, planets and stars, although he did not distinguish the differences between them. His order for closest to farthest from Earth read: the Moon, Mercury, Venus, Sun, Mars, Jupiter and Saturn.

The Greek astronomer Ptolemy (centre).

NICOLAUS COPERNICUS

Lifetime: 1473-1543.

Nationality: Polish.

Major Achievements:

- Copernicus realized the Earth-centred or geocentric system dating back to Ptolemy was inaccuarate.
- He devised a new heliocentric (Sun-centred) system. Copernicus stated that the Earth and all the other planets revolved around a stationary central Sun.
- Copernicus' ideas were incorporated in his book *The Revolution of the Heavenly Spheres,* completed in 1530.
- The book was not published until 1543, perhaps just a few days before he died.

The Polish astronomer Nicolaus Copernicus

TYCHO BRAHE

A bronze statue of Tycho Brahe in Prague.

Lifetime: 1546-1601

Nationality: Danish.

Major Achievements:

- Brahe discovered a supernova (luminous star explosion) in Cassiopeia in 1572. It is now called Tycho's Star. He suggested this was a 'fixed' star outside the Solar System.
- Brahe plotted the accurate positions of 780 stars over 20 years.
- The astronomer employed the German Johannes Kepler as his assistant to help him with his studies.
- Brahe's star catalogue was not published until after his death (*see Kepler*).

TABULÆ
RUDOLPHINÆ.
QVIBVS ASTRONOMICÆ SCIENTIÆ, TEMPO-
rum longinquitate collapsæ RESTAURATIO continetur;
A Phœnice illo Astronomorum
TYCHONE
Ex Illustri & Generosa BRAHEORUM in Regno DANIÆ
familiâ oriundo Equite,
PRIMUM ANIMO CONCEPTA ET DESTINATA ANNO
CHRISTI MDLXIV: EXINDE OBSERVATIONIBUS SIDERUM ACCURA-
TISSIMIS, POST ANNUM PRÆCIPUE MDLXXII, QUO SIDUS IN CASSIOPEIÆ
CONSTELLATIONE NOVUM EFFULSIT, SERIÒ AFFECTATA...
TANDEM TRADUCTA IN GERMANIAM, INQUE AULAM ET
Nomen RUDOLPHI IMP. anno MDIIC.

TABULAS IPSAS, JAM ET NUNCUPATAS, ET AFFECTAS, SED
MORTE AUTHORIS SUI ANNO MDCI DESERTAS,
JUSSU ET STIPENDIIS FRETUS TRIUM IMPPP.
RUDOLPHI, MATTHIAE, FERDINANDI,
...
IOANNES KEPLERUS,
TYCHONI primùm à RUDOLPHO II. Imp. adjunctus calculi minister; indeq;
trium ordine Imppp. Mathematicus:
Qui idem de speciali mandato FERDINANDI II. IMP.
...
Opus hoc ad usus præsentium & posteritatis, typis, numericis propriis, cæteris
& prælo JONÆ SAURII, Reip. Ulmanæ Typographi, in publicum
extulit, & Typographicis operis ULMÆ curator affuit.

Cum Privilegiis Imp. & Regum Rerúmq; publ. vivo TYCHONI ejúsq; Hæredibus,
& speciali Imperatorio, ipsi KEPLERO concesso, ad anno XXX.
ANNO MDCXXVII.

The title page and an illustration from Brahe's *Rudolphine Tables*, completed by Kepler.

JOHANNES KEPLER

Lifetime: 1571-1630

Nationality: German.

Major Achievements:

- Kepler became Brahe's assistant in 1600.
- He devised the laws of planetary motion, linking a planet's orbit and speed to the Sun.
- The astronomer wrote the first astronomy textbook, *Epitome Astronomiae Copernicanae* (Epitome of Copernican Astronomy). He also wrote a science fiction book called Somnium (Dream).
- Kepler completed and published Brahe's star catalogue, *Rudolphine Tables,* in 1627.

The German astronomer Johannes Kepler.

ASTRONOMICAL DISCOVERIES TIMELINE

1619
Kepler's third law of planetary motion published.

1632
An official observatory was set up in Leiden, Netherlands.

1668
Isaac Newton builds the first reflecting telescope in England.

1672
Britain's Royal Observatory established at Greenwich, London.

1675
Danish astronomer Ole Romer measures the speed of light.

1687
Newton published *Principia,* explaining laws of motion and gravitation.

1705
Halley correctly predicts comet seen in 1682 would return in 1758.

1781
William Herschel discovers Uranus, seventh planet of the Solar System.

French astronomer Messier discovers galaxies, nebula and star clusters while looking for comets. He compiles a catalogue of these objects.

1796
Laplace publishes his theory of the origin of the Solar System.

1801
Giuseppe Piazzi discovers the first asteroid, Ceres.

1843
German astronomer Samuel Heinrich Schwabe describes the sunspot cycle.

1860
Sir William Huggins begins the first proper spectral analysis of stars.

1923
Hubble showed that there were galaxies outside our own galaxy, moving apart.

1937
First radio telescope built.

Astronomy
General study of objects in space including stars, planets, moons, galaxies and others. Often includes observing and recording.

Astrophysics
The physical nature of stars, planets and other space bodies, including their make-up and contents, temperatures and pressures, densities and conditions.

Cosmology
Study of the origins, history, make-up and fate of the Universe as a whole, often carried out using mathematics and physics, rather than stargazing.

Space science
Often more concerned with spacecraft, probes, rockets and other hardware, and the conditions for space travellers.

LATER ASTRONOMERS

After the telescope was invented, many more people began gazing up at the night sky. Some had little science background, did it for a relaxing hobby, but chanced upon an amazing discovery that put their name forever into history. Others were full-time professional astronomers who spent a lifetime observing and recording, yet their names are known to very few. This element of chance is smaller today. But it still exists, and draws millions of people to watch the skies every night.

GALILEO GALILEI

Lifetime: 1564-1642

Nationality: Italian.

Major Achievements:

- Galileo improved the first telescopes and was the first person to use them for scientific studies of the night skies.
- He observed mountains and craters on the Moon, many stars too faint to see with the unaided eye, and four moons going around Jupiter.
- Galileo recorded his early discoveries in his book *Sidereal Messenger* (1610).
- Galileo believed the ideas of Copernicus – that the Sun and not the Earth was the centre of the Solar System as had been previously stated by Ptolemy.
- He put forward both sets of theories in his book *Dialogue on Two Chief World Systems* (1632). This work was heavily criticised and the astronomer was put under house arrest by religious leaders for his views.
- Galileo made advances in many other areas of science including the mechanics of moving objects like swinging pendulums, falling canonballs and bullets from guns.

• See page 8 for information on Ptolemy.

GIOVANNI DOMENICO CASSINI

Lifetime: 1625-1712

Nationality: Italian-French.

Major Achievements:

- Cassini was appointed as Director of the Paris Observatory in 1669.
- He made many discoveries including four satellites of Saturn and the gap in Saturn's rings (now named the Cassini Division).
- Cassini made many advances combining his observations with calculations, including the orbit times of Mars, Venus and Jupiter, the paths of Jupiter's moons, and the first fairly accurate measurement from the Earth to the Sun (the AU, Astronomical Unit).

JOHN FLAMSTEED

Lifetime: 1646-1719

Nationality: British.

Major Achievements:

- Flamsteed became the first Astronomer Royal, in 1675.
- He made the first extensive star charts using the telescope, as part of work aimed at giving sailors a better method of navigation. The charts recorded positions of over 2,935 stars.
- Due to a dispute with Isaac Newton and the Royal Society the charts were not published until six years after Flamsteed's death.

EDMOND HALLEY

Lifetime: 1656-1742

Nationality: British.

Major Achievements:

- Edmond Halley travelled to St Helena in the South Atlantic at the age of 20 to make the first telescopic chart of stars as seen in the Southern Hemisphere.
- Halley became interested in comets after the 'Great Comet' of 1680. He worked out from historical records that a comet seen in 1531, 1607 and 1682 should return in 1758, which it did. The comet is now named after Halley.
- The astronomer was the first to suggest that nebulae were clouds of dust and gas inside which stars might form.
- Halley became Astronomer Royal in 1720, and began an 18-year study of the complete revolution of the Moon.
- Halley's other activities included studying archaeology, geophysics, the history of astronomy.

WILLIAM HERSCHEL

Lifetime: 1738-1822

Nationality: German-British.

Major Achievements:

- Herschel made many of his own telescopes.
- He discovered the planet Uranus in 1781 and some moons of Uranus and Saturn.
- During his lifetime, Herschel catalogued over 800 double-stars.
- He also published a chart of over 5,000 nebulae in 1820.
- Herschel also recognised that the Milky Way was a flattened disk of stars.

EDWIN HUBBLE

Lifetime: 1889-1953

Nationality: American.

Major Achievements:

- Working mostly at Mount Wilson Observatory, Hubble's studies of nebulae such as parts of Andromeda showed they were masses of stars.
- Hubble concluded that these star masses were galaxies outside our own Milky Way.
- He introduced a system of classifying galaxies by their shapes.
- Hubble measured the speed of galaxies in 1929 and showed farther ones move faster, leading to Hubble's Law and the idea that the Universe is expanding.

PERCIVAL LOWELL

Lifetime: 1855-1916

Nationality: American.

Major Achievements:

- Lowell became interested in astronomy after reports by Schiaparelli of 'channels' on Mars. 'Channels' was misunderstood as 'canals', and Lowell became convinced of the existence of Martians, even writing books on them.
- He established the Lowell Observatory on a 2,200-metre peak in Arizona in 1894, mainly to study Mars.
- Lowell predicted the existence of another planet beyond Neptune. This planet, Pluto, was discovered in 1930 at Lowell's observatory.

STEPHEN HAWKING

Lifetime: 1942 -

Nationality: British.

Major Achievements:

- Hawking continued Einstein's ideas on time being a fourth dimension, and worked on the origin of the Universe at the Big Bang.
- He worked on a common theory for the four basic forces in the Universe – gravity, electromagnetic, strong nuclear and weak nuclear forces.
- Hawking made great advances to our understanding of black holes, detailed in his book *A Brief History of Time*.

ASTRONOMICAL DISCOVERIES TIMELINE

1948
Five-metre Hale reflector telescope came into use at Mount Palomar, California, USA.

1950
Jan Oort predicts existence of Oort Cloud of comets.

1951
Gerard Kuiper proposes existence of Kuiper Belt of comets.

1962
First X-rays detected from space.

1963
First quasar (quasi-stellar object) discovered.

1967
First pulsar (spinning neutron star) discovered.

1976
Six-metre reflector telescope came into use at Mount Semirodniki, USSR.

1986
Halley's comet passes Earth.

1987
SN1987A became the first supernova to be seen with the unaided eye in modern times.

1990
Hubble Space Telescope sent into Earth orbit by the Space Shuttle.

1992
COBE satellite detected microwave 'echoes' of the Big Bang.

1998
Astronomers discover that the expansion of the Universe appears to be speeding up. It will expand forever.

2003
WMAP satelllite makes first detailed map of minute temperature variations in the microwave echoes of the Big Bang.

2005
Astronomers announce discovery of 2003UB313, the largest object to be found in the outer Solar System since Pluto in 1930.

THE SUN

Our closest star, the Sun is the centre of the Solar System. All the planets and asteroids are held in their orbits by its immense gravity. It also attracts visitors from the farthest reaches of the system, such as comets. For billions of years the Sun has been providing Earth with light that green plants use as an energy source for living and growing. Herbivorous animals eat the plants, carnivorous animals eat the herbivores, and in this way the Sun powers life on Earth – including our own.

WHERE IN THE SOLAR SYSTEM?

ORBIT DETAILS

Average distance from centre of Milky Way 26,000 light years (250,000 million million km)

Time for one orbit around centre 225 million years

Average orbital speed 217 km per second

Time to spin on axis 25.38 days at equator

STAR PROFILE

Diameter at Equator	**1,392,000 km**
Surface area	**6.10 million million sq km**
Mass	**1.99 billion billion billion tonnes**
Volume	**1.41 billion billion cu km**
Overall density	**1.408 g per cc**
Gravity (Earth = 1)	**27.9**
Number of main planets	**9 (debated)**

A NASA photograph of the Sun.

STRUCTURE AND LAYERS

CORE
- About 280,000 km across.
- Nuclear fusion reactions convert hydrogen to helium, producing immense amounts of light, heat and other radiation.
- Energy output equivalent to 90 million billion tonnes of TNT per second.

RADIATIVE ZONE
- About 350,000 km deep.
- Conveys heat and light outwards by photon transfer between ions.
- Temperature falls with distance to core.

INTERFACE LAYER
- A thin layer between the radiative zone and the convective zone.

CONVECTIVE ZONE
- About 200,000 km deep.
- Super-hot material carries heat outwards from radiative zone.
- Material cools at photosphere and sinks back to receive more heat, causing convection currents.

PHOTOSPHERE
- Visible surface of the Sun.
- Around 600 km deep.
- Emits photons of light and other energy forms into space.

CHROMOSPHERE
- About 2,500 km deep.
- Visible as a red-coloured flash around the Sun at the start and end of a total solar eclipse.

CORONA
- Wispy outer atmosphere around the Sun.
- Extends many millions of km into space, to distances bigger than the Sun itself.

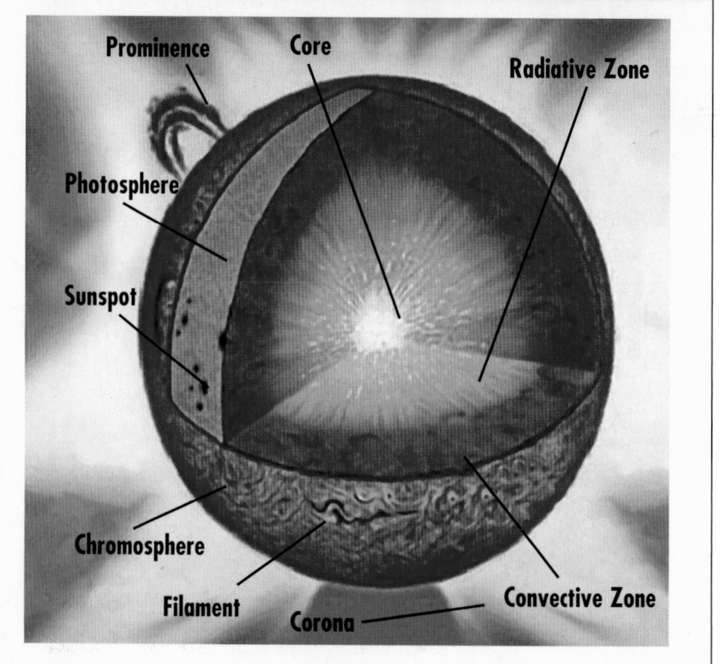

FLARES AND PROMINENCES

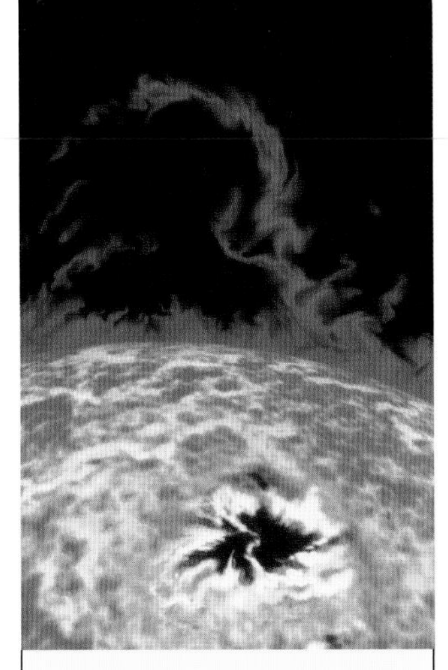

- Solar flares are massive explosions in the lower corona and chromosphere. They were first observed in 1859.
- They trigger massive solar eruptions called coronal mass ejections.
- Solar prominences are larger and longer-lasting than flares.
- Many leap up, along and down in a curved arc back to the Sun.
- Typically, prominences are thousands of km long.
- The largest ones can be over 500,000 km long.

MAKE-UP OF PHOTOSPHERE

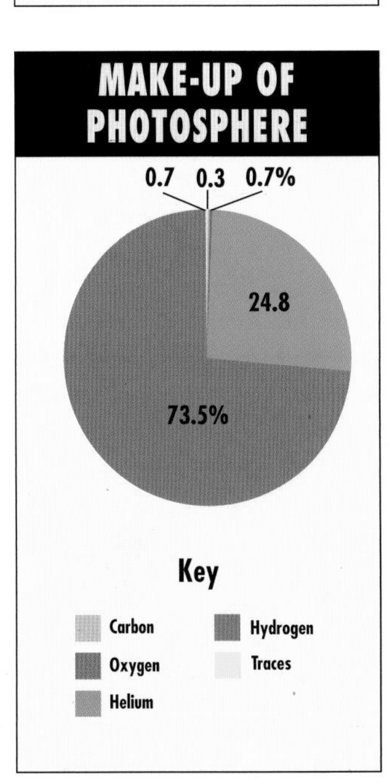

SOLAR WIND

- Solar wind steams away from the Sun in all directions.
- It reaches speeds of up to 400 km per second and comes mainly from the corona.
- Solar wind consists of charged particles, ions and other particles in a form called plasma.
- Where solar wind interacts with Earth's magnetic fields, near the North and South Poles, it creates aurora, shimmering 'curtains' of light high in the sky, the Northern Lights (*Aurora Borealis*) and Southern Lights (*Aurora Australis*).

• See page 54 for more information on solar wind.

A diagram of solar wind. The Earth (right) is protected by its magnetic field.

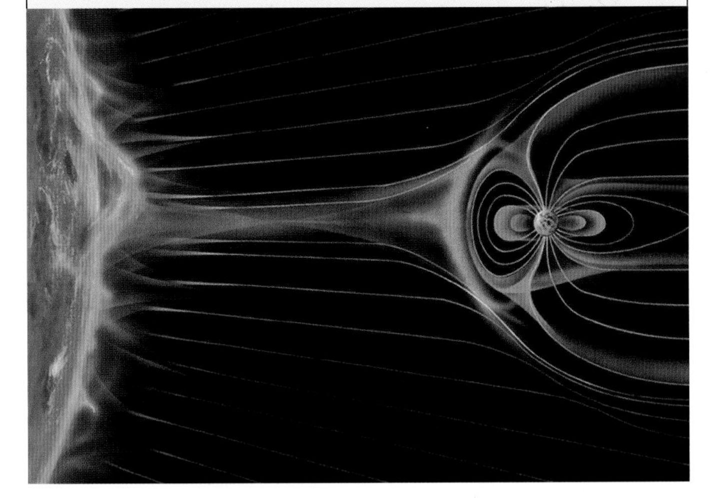

SUNSPOTS

Sunspots are cooler variable patches on the photosphere, probably caused by magnetic interactions.

- The inner umbra of each spot is around 4,000°C. The outer penumbra is about 5,500°C.
- They were first noticed to vary in a regular or cyclic way by Heinrich Schwabe between 1826 and 1843.
- Sunspots usually vary in an 11-year cycle. An average sunspot's 'life' is 2 weeks.
- On 30 March 2001 SOHO (Solar and Heliospheric Observatory) recorded the largest sunspot group so far, covering more than 13 times the area of the Earth.

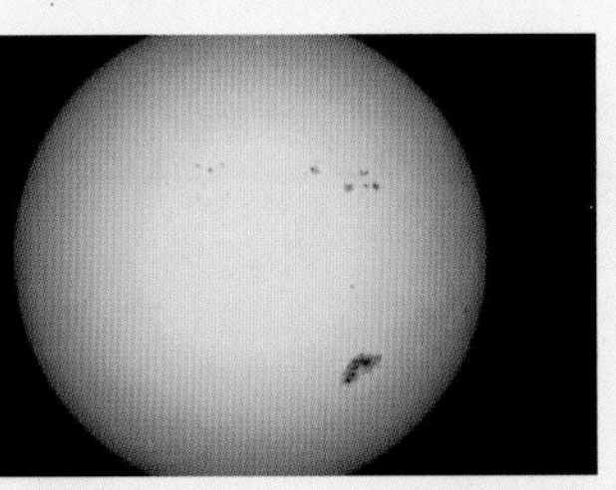

NASA photo of a sunspot.

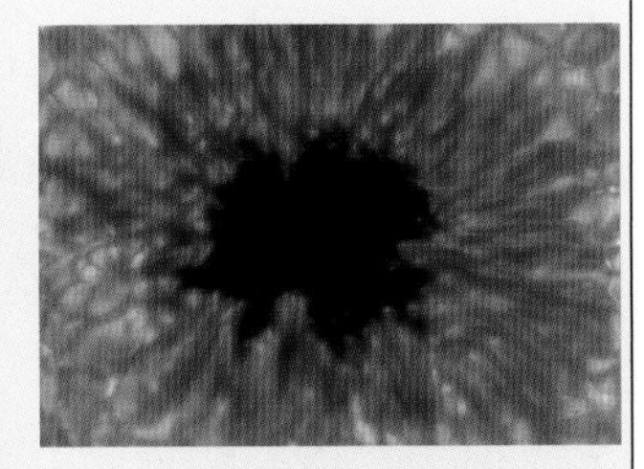

Close-up of a sunspot.

TEMPERATURES

Corona 2 million °C **Surface** 5,800 °C **Core** 13.5 million °C

SUN TIMELINE

27,000 years ago
Depicted in rock carvings in Europe, North Africa, Australia and cave paintings 10,000 years later.

From 7,000 years ago
Sun worshipped as a god by many ancient civilizations.

4,900 years ago
First phase of construction of Stonehenge, a Sun-aligned stone-age temple in England.

From 4,000 years ago
The Sun worshipped as the god Ra in ancient Egypt.

2,030 years ago
Chinese astronomers first mentioned sunspots.

1300s
Aztec people made sacrifices to their Sun god Huitzilopochtli.

1610
Sunspots first seen through a telescope by Johannes and David Fabricius, then by Galileo.

1962
McMath Pierce Solar Telescope erected in Arizona, USA, the largest telescope dedicated to Sun study.

1990
Ulysses probe launched from Space Shuttle to study the Sun's North and South Poles and solar wind.

1995
Joint European/US probe SOHO launched on 2 December.

1997
ACE (Advanced Composition Explorer) satellite launched to study particles and materials from the Sun and elsewhere.

2001
Space probe Genesis was launched on 8 August to capture samples of solar wind.

2004
On 8 September Genesis returned but was damaged crash-landing.

2004
Analysis begins on the hundreds of Sun samples retrieved from the Genesis probe.

MERCURY

Known by most ancient people from its brief periods of visibility at dawn and dusk, Mercury was named after the Roman winged messenger of the gods – because it has the fastest orbital speed of any planet, averaging 47 kilometres every second. Being the closest planet to the Sun, it is blasted by solar heat and other radiation. This has swept away all but the flimsiest atmosphere, and heats Mercury's daytime side to incredible temperatures – yet the night side plunges to within 100°C of absolute zero.

WHERE IN THE SOLAR SYSTEM?

SURFACE CONDITIONS

ATMOSPHERE:
Almost zero – traces of potassium, argon, oxygen, argon.

NATURE OF SURFACE:
Bare iron-rich rocks pitted with hundreds of large craters.

AVERAGE SURFACE TEMPERATURE: 170°C

LOWEST SURFACE TEMPERATURE: Minus 183°C

HIGHEST SURFACE TEMPERATURE: 425°C

WEATHER OR CLIMATE:
None due to lack of atmosphere.

SEASONAL CHANGES:
None due to almost zero tilt of axis.

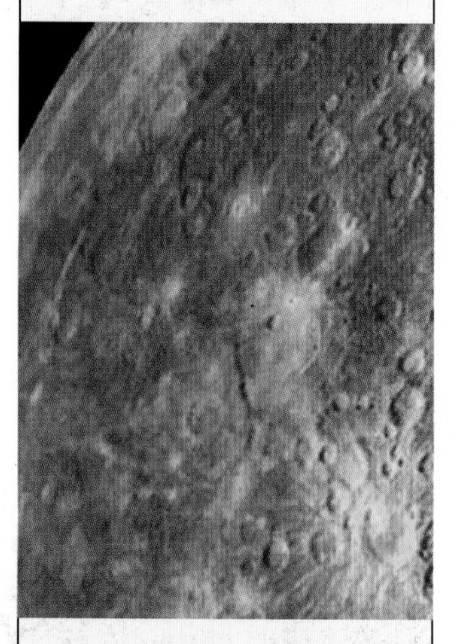

A colour photograph of Mercury, showing the pitted iron-rich surface.

PLANET PROFILE

Diameter at Equator **4,879 km**

Surface area **75 million sq km**

Tilt of axis **0.01°**

Mass (Earth = 1) **0.055**

Volume (Earth = 1) **0.056**

Overall density **5.42 g per cc**

Gravity (Earth = 1) **0.377**

Number of moons **0**

A NASA photograph of the planet Mercury.

ORBIT DETAILS

Average distance from Sun 57.9 million km

Average distance from Sun 0.387 AU (Earth =1)

Closest distance to Sun (perihelion) 46.0 million km

Farthest distance from Sun (aphelion) 69.8 million km

Average orbital speed 47.4 km per second

Slowest orbital speed 38.9 km per second

Fastest orbital speed 59.0 km per second

Time for one orbit (Mercury year) 87.9 Earth days

Axial rotation period 58.6 Earth days

Length of Mercury day 176 Earth days

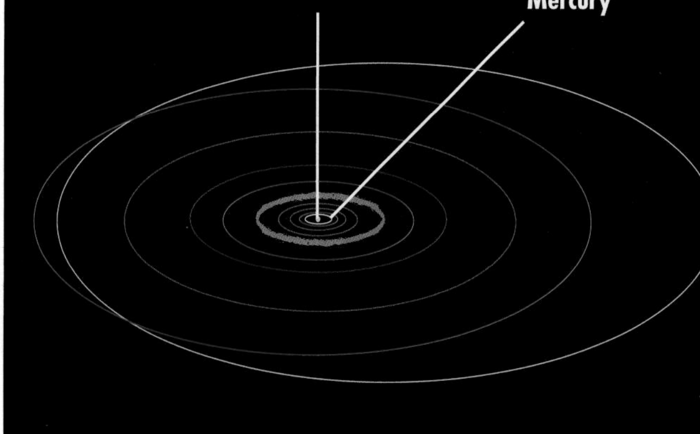

MAJOR FEATURES

Less than half of Mercury's surface has been mapped in any detail (by Mariner 10) so its surface features are less known than most other planets.

Caloris Basin
Massive crater made by asteroid/meteoroid impact, measuring 1,350 km across.

Caloris Montes
'Hot Mountains', curved ranges with peaks rising to 3,000 m sited at one of the hottest places on Mercury, within the Caloris Basin crater.

Discovery Scarp
Joining two craters, this 'cliff' is 350 km long and its maximum height is around 2,800 m.

The heavily pitted Caloris Basin crater.

• See pages 34-37 for information on asteroids and meteors.

OTHER FEATURES

- **SCARPS (RUPES)** Long cliff-like ridges with one steep side and one gradually sloping side, formed as below.
- **RIDGES (DORSA)** Long prominent ridges with two steep sides, formed as Mercury's core cooled and shrank and the already-solid crust was cracked into wrinkles.
- **YOUNGER PLAINS** Uplands probably formed from hardened lava flows, less marked by craters from impacts.
- **OLDER PLAINS** Lowlands much more pockmarked with overlapping craters than the younger plains.
- **ARECIBO VALLIS** Valley named after the Arecibo Observatory, home of Earth's largest radio telescope, in Puerto Rico.
- **ICE** Despite Mercury's incredible heat, there is probably ice (frozen water) at its North Pole, in deep craters with permanent shade from the Sun.

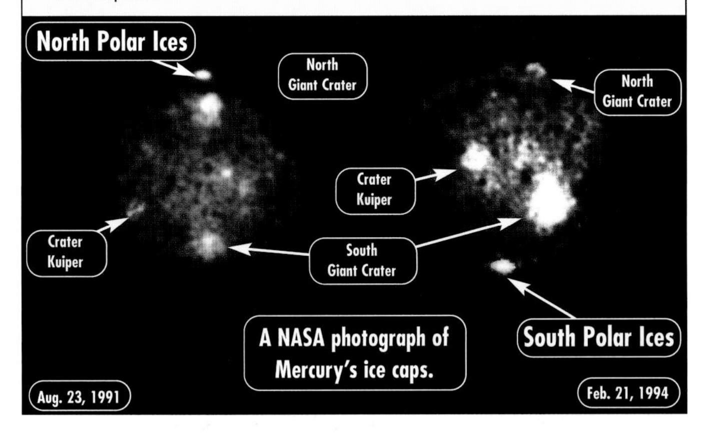

A NASA photograph of Mercury's ice caps.

SMALL AND CURIOUS

The minute size of Mercury is shown in the above picture of the planet (circled in red) in transit against the sun.

- Mercury is the second-smallest planet in the Solar System, after Pluto.
- It has a very 'eccentric' or oval-shaped orbit, much more so than most other planets – again, only outermost Pluto is more 'eccentric'.
- Its axis is hardly tilted at all, being at right angles to the Sun, so the Sun is always directly over its Equator all through its year.

DISTINGUISHING FEATURES

Mercury has several features that mark it out from other planets.

Mercury has several hundred named craters, with names like Dickens, Shakespeare, Chopin, Mark Twain, Beethoven, Degas and Sibelius. All of its craters are named after famous artists and classical musicians.

In 1974, the Mariner 10 spacecraft produced this image of the 27 mile wide Degas crater.

TEMPERATURE

Mercury has the widest temperature range of any planet, spanning almost 600°C between day on the sunny side and night on the shady side. (Earth's maximum range is less than 150°C.)

SPINNING MERCURY

Because of its closeness to the Sun and slow spinning speed, at certain places and times on Mercury an astronaut would see the Sun rise to just above the horizon, then go back and set, and then rise again – all on the same Mercury day.

MERCURY TRANSIT

Since Mercury is nearer the Sun than Earth, when the two planets are almost in line, Mercury appears to cross the disc of the Sun when viewed from Earth. This is called the transit of Mercury. There are about 14 transists every century.

MERCURY TIMELINE

5,000 years ago
The Sumerians mentioned Mercury, which they call Ubu-idim-gud-ud.

3,300 years ago
Earliest detailed observations of Mercury in ancient Babylon.

2,500 years ago
In ancient Greece, Mercury (like Venus) was though to be two different planets with two names – Apollo in the dawn sky and Hermes at dusk.

2,470 years ago
Heraclitus thought that Mercury, along with Venus, orbited the Sun rather than the Earth.

1,000 years ago
Ancient Chinese documents refer to Mercury as the 'Water Star'.

1639
Giovanni Zupi's telescope observations showed different parts of Mercury were lit at different times by the Sun.

1961-62
The Mercury program of single-seat craft was the first to carry US astronauts into space.

1965
Radar measurements showed that Mercury does not spin once but three times for every two orbits.

1973
US Mariner 10 launched in November to fly past Venus and Mercury.

1974
Mariner 10 makes first flypast of Mercury on 29th March.

1975
Mariner 10's third flypast in March returns information on Mercury's magnetic field.

2004
US Mercury probe MESSENGER launched in August.

2008-09
MESSENGER due to make three flybys of Mercury.

2011
MESSENGER due to enter Mercury orbit in March and survive for a year, studying the thin atmosphere and explore the surface in detail.

VENUS

Second planet from the Sun, Venus is named after the Roman goddess of love – and is shrouded in mystery. In fact, it is shrouded by thick swirling clouds of gases and droplets of acid that hide its surface from the gaze of outsiders. Although Venus is about the same size and mass ('weight') as Earth, it could not be more different. It is the hottest of all the planets, partly because its thick atmosphere traps vast amounts of heat from the nearby Sun in a 'runaway greenhouse effect' far more extreme than here on Earth.

WHERE IN THE SOLAR SYSTEM?

SURFACE CONDITIONS

ATMOSPHERE:
Thick, dense, mainly carbon dioxide, also nitrogen and sulphur acids.

NATURE OF SURFACE:
Hard and rocky, with numerous volcanoes.

AVERAGE SURFACE TEMPERATURE:
470°C

LOWEST SURFACE TEMPERATURE:
45°C (at cloud tops)

HIGHEST SURFACE TEMPERATURE:
500°C in valleys near the equator.

WEATHER OR CLIMATE:
Thick swirling deadly-poisonous atmosphere. Winds reach 300 km/h at the top of the clouds.

SEASONAL CHANGES:
Minimal on surface due to dense atmosphere.

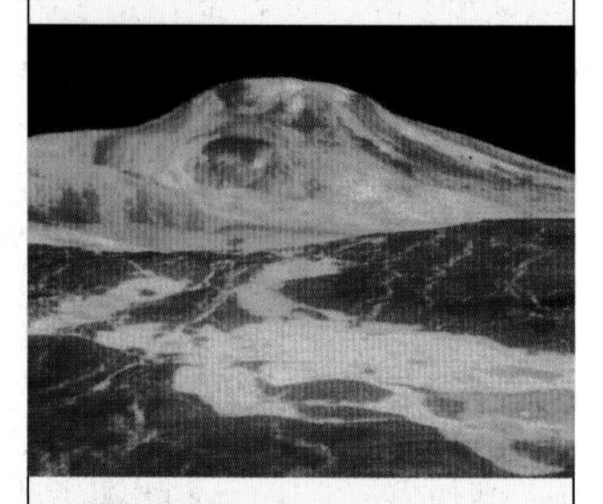

A Magellan radar image of the volcanoo *Sif Mons* on Venus.

PLANET PROFILE

Diameter at Equator **12,103 km**

Surface area **460 million sq km**

Tilt of axis **177.3°**

Mass (Earth = 1) **0.815**

Volume (Earth = 1) **0.856**

Overall density **5.2 g per cc**

Gravity (Earth = 1) **0.90**

Number of moons **0**

A NASA mosaic of the planet Venus.

ORBIT DETAILS

Average distance from Sun
108.2 million km

Average distance from Sun
0.723 AU (Earth =1)

Closest distance to Sun (perihelion) 107.4 million km

Farthest distance from Sun (aphelion) 108.9 million km

Average orbital speed
35.0 m per second

Slowest orbital speed
34.8 m per second

Fastest orbital speed
35.3 km per second

Time for one orbit
(Mercury year) 224.7 Earth days

Axial rotaton period
243 Earth days

Length of Venus day
117 Earth days

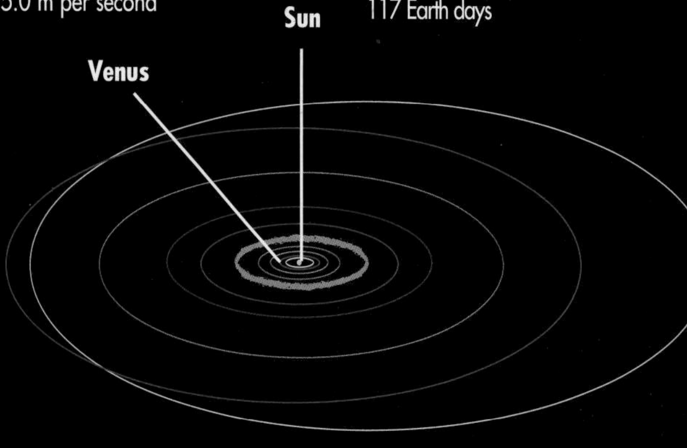

MAJOR FEATURES

Several probes have been sent to Venus, and radio waves have been used to map virtually the entire planet.

Ishtar Terra
Northern highlands about the size of Australia, bearing Venus's highest mountains.

Maxwell Montes
A range of mountains about 850 km long, with the highest peaks over 11,000 m tall.

Lakshmi Planum
Vast upland plain or plateau partly encircled by Maxwell Montes.

Aphrodite Terra
Southern uplands, roughly the size of South America.

Arachnoid Volcanoes
Photographed by the space probe Magellan, these have unusual ridges around them. The central volcano with its surrounding ridges looks like a giant spider.

A NASA photograph of the sprawling Aphrodite Terra, shown in brown.

OTHER FEATURES

- **CORONAE** Circular centres surrounded by ring-like ridges, the largest being Artemis Corona at 2,100 km across.
- **PLAINS** Flat and fairly smooth, these cover two-thirds of the surface with low volcanoes up to 200 km across.
- **MOUNTAINS** Six main mountain ranges covering about one-third of the surface.
- **UPLAND REGIONS** One of the largest is Beta Regio, dominated by Rhea Mons and Theia Mons.
- **LOWLAND DEPRESSIONS** Wide and low, these include Atalanta Planitia, Guinevere Planitia, and Lavinia Planitia.
- **ALL FEATURES** All of Venus's surface features are named after females (either real people or from myth and legend), with the exception of the Maxwell Montes, named after scientist James Clerk Maxwell.

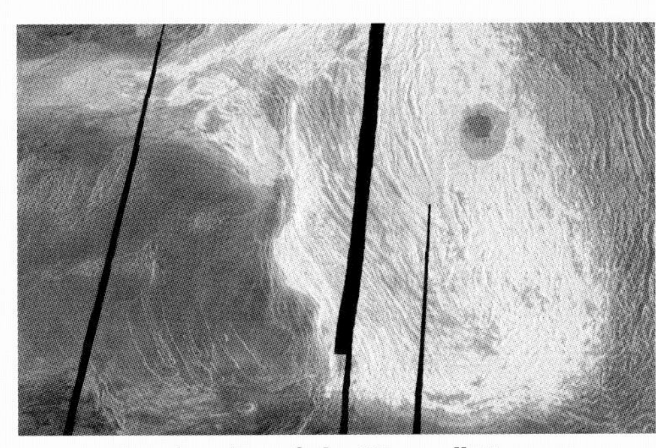

A radar shot of the Maxwell Montes.

TRANSIT OF VENUS

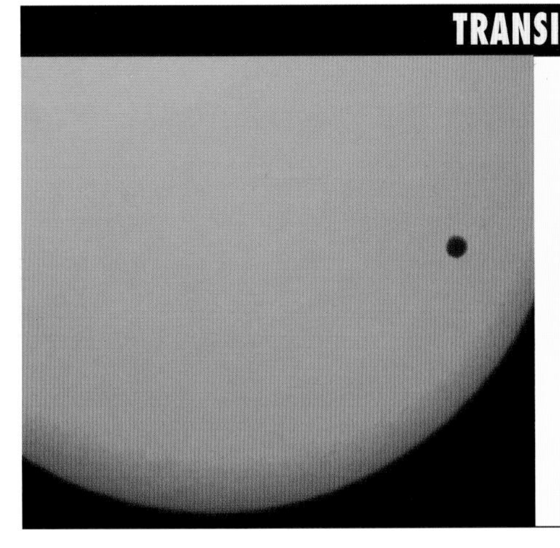

Since Venus is nearer the Sun than Earth, when the two planets are almost in line, Venus appears to cross the disc of the Sun when viewed from Earth. This is called the transit of Venus.The date it occurs and the time Venus takes to cross the Sun's face have been used to estimate the distance between Earth and Sun. Transits occur in pairs. The two in each pair are about eight years apart, but the time between pairs is more than 100 years.

The transit of Venus across the Sun.

• *See page 18 for information on the Earth's orbit.*

DAYTIME VIEWING

- Because of its closeness and bright reflection of sunlight, Venus is so bright that it is one of only two space bodies, other than the Sun, which can be seen during daylight from Earth – the other being the Moon.
- It is also often the first star-like body to appear at dusk and the last to fade at dawn, earning it the names 'Evening Star' and 'Morning Star'.

REVERSE SPIN

Venus is one of only three planets with retrograde spin (the others are Uranus and Pluto). This means it spins on its axis in the opposite direction to the other planets. Seen from the side, its surface moves from east to west or right to left, or clockwise if viewed from above its North Pole.

NEAREST NEIGHBOUR

No other planet comes closer to the Earth than Venus. At its closest it is 38.2 million km away.

SLOW COACH

Venus takes longer to spin once on its axis than it takes to comoplete one orbit of the Sun.

CIRCULAR ORBIT

Most planets have an orbit which is an ellipse (oval). The journey of Venus around the Sun is the most circular of all planets – that is, it is the least 'eccentric' orbit (especially compared to that of Venus' close neighbour Mercury).

UNDER PRESSURE

The atmosphere's pressing force or pressure on Venus (pictured below next to the Earth) is incredible – 90 times more than our own, and equivalent to the pressure almost 1,000 metres under the sea here on Earth.

VENUS TIMELINE

3,600 years ago
Astronomical records in Babylonia record appearances of Venus.

3,500 years ago
Ancient Babylonians record Venus as one of the brightest 'stars'

2,500 years ago
In ancient Greece, Venus was though to be two different planets with two names – Phosphorus in the dawn sky and Hesperus at dusk.

2,000 years ago
Ancient Chinese observers refer to Venus as the 'Metal Star'.

1610
Galileo observed phases of Venus.

1639
First transit of Venus was observed.

1672
Giovanni Domenico Cassini claimed to discover a moon of Venus.

1961
Russian space probe Venera 1 aimed at Venus, but failed.

1962
US probe Mariner 2 flew past Venus.

1966
Venera 3 probe crash-landed on the surface. Venera 4 was more successful next year and sent back information. Veneras 5, 6 and 7 sent back information in subsequent years.

1970
Venera 7 made the first successful landing.

1975
Venera 9 was the first spacecraft to send a picture back from the surface of Venus on 21 October.

1978
US sent two Pioneer probes to Venus.

1990
The Magellan probe mapped all but 1/50th of the surface.

1998-99
Cassini-Huygens flew past Venus.

2004
First of a pair of transits witnessed.

2006
European Space Agency's Venus Express orbiter probe is due to arrive.

2012
Next transit of Venus due.

EARTH

We think of our home world as 'average' for a planet. But the more we learn about the rest of the Solar System, the more we see that Earth is very unusual. Earth is the only planet with abundant water in liquid form. This is mainly because its average surface temperature is just 22 degrees centigrade. Indeed, more than three-quarters of Earth's surface is rivers, lakes, seas, oceans, and frozen water as glaciers and ice-caps.

WHERE IN THE SOLAR SYSTEM?

SURFACE CONDITIONS

ATMOSPHERE: Almost four-fifth nitrogen, one-fifth oxygen, traces of carbon dioxide, water vapour and other gases.

NATURE OF SURFACE: Varied from high rocky mountains to deep valleys and trenches, mostly covered with water.

AVERAGE SURFACE TEMPERATURE: 22°C

LOWEST SURFACE TEMPERATURE: minus 89°C

HIGHEST SURFACE TEMPERATURE: 58°C

WEATHER OR CLIMATE: Varies due to movement of atmosphere and its water vapour distributed by clouds and falling as rain. Generally conditions become colder from the Equator (tropics) to the Poles.

SEASONAL CHANGES: Marked seasons due to considerable tilt of axis, cold icy winters and hot summers.

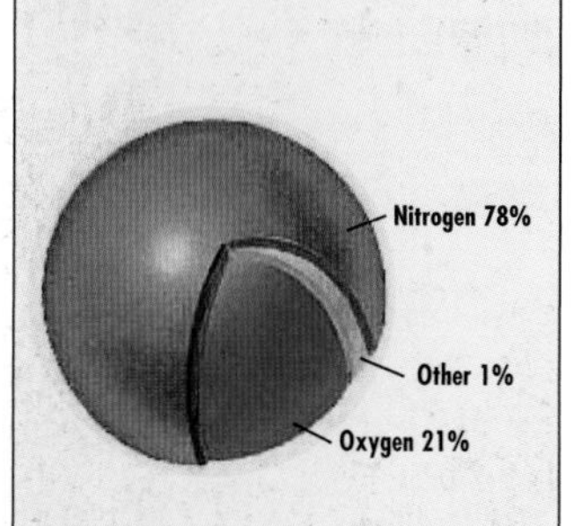

PLANET PROFILE

Diameter at Equator **12,726 km**

Surface area **510 million sq km**

Tilt of axis **23.4°**

Mass **5,974 billion billion tonnes**

Volume **1,083 billion cu km**

Overall density **5.517 g per cc**

Gravity **1g (9.78 m per s per s)**

Number of moons **1**

A NASA photograph of Earth.

ORBIT DETAILS

Average distance from Sun 149.6 million km

Average distance from Sun 1.0 AU

Closest distance to Sun (perihelion) 147.1 million km

Farthest distance from Sun (aphelion) 152.1 million km

Average orbital speed 29.8 km per second

Slowest orbital speed 29.3 km per second

Fastest orbital speed 30.3 km per second

Time for one orbit (Earth year) 365.256 Earth days

Axial rotation period (Earth day) 23.93 Earth hours

Sun

Earth

MAJOR FEATURES

The Earth has been mapped extensively.

Rivers
The River Amazon of South America carries more water than the next five biggest rivers combined, emptying 200,000 cubic litres per second into the Atlantic Ocean.

Oceans
The Pacific Ocean covers almost half (46%) of the Earth's surface.

Mountains
The Himalayas of central Asia and northern India have eight of the world's ten tallest peaks.

Deserts
The Sahara Desert of North Africa is by far the greatest arid (very dry) area, covering more than 9 million sq km.

Lowest Point
The bottom of the deep-sea Marianas Trench in the north-west Pacific Ocean is 10,911 m below the ocean's surface.

Highest Point
The peak of Mount Everest in the Himalayas is 8,848 m above sea level.

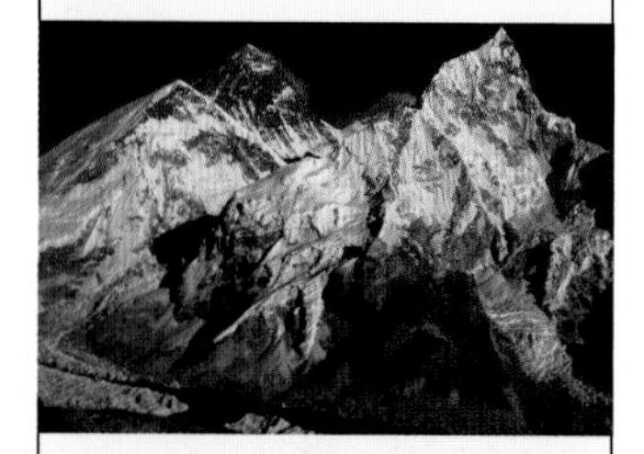

Earth's highest mountain, Everest.

OTHER FEATURES

The outer reef of the Great Barrier Reef.

- **LAKE SUPERIOR** The largest body of fresh water by area on Earth.
- **LAKE BAIKAL** The largest body of fresh water by volume on Earth.
- **THE LAMBERT GLACIER** Situated on Antarctica, it is the largest glacier on Earth, 500 km long and 80 km wide.
- **THE GRAND CANYON** This spectacular deep valley has been worn away by the Colorado River. It is 350 km long, up to 30 km wide, and in places 1,600 m deep between almost sheer cliffs.
- **THE GREAT BARRIER REEF** A long series of rocky reefs built over thousands of years by billions of tiny animals called coral polyps.

PLATE TECTONICS

- The Earth's outer surface of thin rocky crust, plus a thin layer of semi-melted rock below, is split into 12-15 giant curved pieces called lithospheric plates.
- Over millions of years these slide or drift around the globe, at the rate of about 1-3 cm per year, carrying the major land masses with them in a process called continental drift.
- At the edges of some plates new rock is added, by a process called seafloor spreading.
- Where two plates ram into each other, the crust buckles into mountains, such as the Himalayas and Andes.
- Where one plate slides below the other (subduction zone) there are earthquakes and volcanoes.

POLAR ICE

- Apart from large cloud systems, the glistening ice caps over the North and South Poles are perhaps Earth's most noticeable feature from space.
- Each shrinks in summer, then spreads in winter, due to Earth's seasonal changes and zoned climate from Equator to Pole.
- The Arctic ice cap over the North Pole is a 'raft' of ice up to 10 m thick floating in the Arctic Ocean, with a winter extent of 15 million sq km.
- The Antarctic ice cap over the South Pole covers the vast southern land mass of Antarctica, with a winter extent of 19 million sq km

LIFE

- Earth is the only body in the Solar System known to support life. This life depends on liquid water, and mostly occurs in the narrow temperature range of 0 to 40°C.
- The greatest variety of land life occurs in the tropical rainforests, which have 9 out of 10 of the 20-plus million species, of plants, animals and other life forms.
- The richest variety of marine life is found in coral reefs.
- Many areas of wildlife are being affected, polluted, and used for agriculture and industry by the dominant life-form on Earth – *Homo sapiens.*

EARTH TIMELINE

Prehistory
More than 10,000 years ago people made maps of their areas carved on stone or ivory, scratched into tablets or woven into hangings.

3,000 years ago
Early Greeks supposed the Earth was a flat disc.

2,500 years ago
In Ancient Greece the idea grew that the world was round or a globe, based on observations such as how the stars vary at different places on Earth.

1519-1522
Ferdinand Magellan's expedition circled the globe to show that Earth was indeed a sphere.

1785
James Hutton proposed his Principle of Uniformitarianism, which means the Earth's surface has been shaped over huge lengths of time by the same processes we see at work today - volcanoes, earthquakes, mountain-building, and erosion by wind, rain, ice and snow. He believed the Earth was 'immeasurably ancient'.

1882
William Thomson (Lord Kelvin) worked out the Earth's age from its cooling rate. His approximate age for the Earth was one-tenth of today's estimate.

1908
Frank Taylor and Howard Baker put forward a scientific explanation for continental drift.

1912
Alfred Wegener suggested a version of the modern theory of plate tectonics, which causes continental drift.

1956
Clair Patterson determined from amounts of radioactivity in rocks that the Earth is 4,500 million years old, today's accepted age being nearer 4,600 million.

1960s
Earth scientists came to accept Weneger's basic ideas and developed the modern version of plate tectonics.

1989
The first of the 24 operational NAVSTAR satellites was launched that from the 1990s would provide the GPS, Global Positioning System, for 'satellite navigation) to locate any spot on Earth's surface with a few tens of metres.

THE MOON

A moon is a natural object of reasonable size going around a planet. The one we call the Moon (usually with 'the' and a capital 'M') is Earth's single moon. It has also been known to scientists as Luna, and the word 'lunar' means to do with the Moon. Seen from Earth, the Moon is about the same-sized disc in the sky as the Sun. It appears to change shape during its 29.5-day orbit because we can only seen the sunlit part of its surface, creating the phases of the Moon. Its pull of gravity also makes the water in seas and oceans rise and fall, which we call the tides.

WHERE IN THE SOLAR SYSTEM?

ORBIT DETAILS

Average distance from Earth 384,400 km

Average distance from Earth 0.0026 AU (Earth = 1)

Closest distance to Earth (perigee) 363,100 km

Farthest distance from Sun (aphelion) 405,700 km

Average orbital speed 1.02 km per second

Slowest orbital speed 0.97 km per second

Fastest orbital speed 1.08 km per second

Time for one orbit (Earth units) 29 days 12 hours 44 minutes

Axial rotation period (Earth units) 27 days 7 hours 43 minutes

MOON PROFILE

Diameter at Equator **3,476 km**

Surface area **37.9 million sq km**

Tilt of axis **3.6 to 6.7°**

Mass (Earth = 1) **0.074**

Volume (Earth = 1) **0.020**

Overall density **3.34 g per cc**

Gravity (Earth = 1) **0.165**

Number of moons **None**

An artist's illustration of the Moon.

LUNATICS

The Moon features greatly in many legends and stories. One superstition was that if a person stared at the Full Moon for too long, he or she would become mad. This is where the word 'lunatic' comes from.

Another legend was that at Full Moon certain people would grow hair, long teeth and claws and become savage and deadly werewolves.

ORIGIN

It's thought the Moon was formed when a huge Mars-sized space lump of rock, provisionally named Theia, crashed into Earth some 4,500 million years ago. This was when the Earth was about 100 million years old. The loose matter and debris orbiting Earth after the impact came together to form the Moon.

MAJOR FEATURES

The moon has been visited several times and mapped.

Largest Crater
The largest known crater in the Solar System, the South Pole-Aitken Basin is 2,250 km across and 13 km deep.

Sea of Tranquility
The site for the first Apollo Moon landing in 1969. 600 by 900 km.

Sea of Serenity
Site for the last Moon landing in 1972, about 550 km wide.

Sea of Crises
Main dark circular area near the upper eastern (right) edge.

Apennine Mountains
Peaks more than 4,500 m high.

Copernicus
Small, 3,000 m deep crater (below).

Ocean of Storms
Largest lowland plain, covering 2.3 million sq km.

The heavily pitted Copernicus crater.

• *See page 55 for information on probes to the Moon.*

SURFACE CONDITIONS

ATMOSPHERE: Tiny traces of helium, neon, hydrogen, argon.

NATURE OF SURFACE: Craters, mountains, valleys, plains called 'seas'.

AVERAGE SURFACE TEMPERATURE: minus 23°C

LOWEST SURFACE TEMPERATURE: minus 230°C

HIGHEST SURFACE TEMPERATURE: 124°C

WEATHER OR CLIMATE: None.

SEASONAL CHANGES: None.

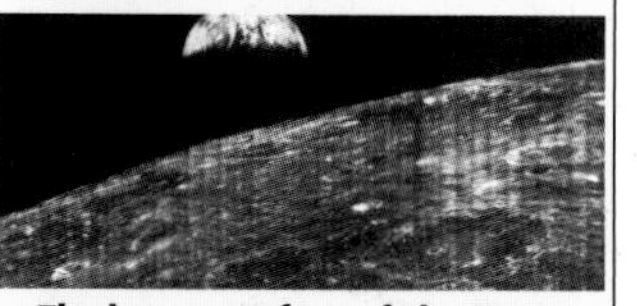

The barren surface of the Moon.

OTHER FEATURES

- **MARIA** Meaning 'seas', these are dark lowland plains of hardened basalt rocks which once flowed as lava. They are totally dry, like the rest of the Moon, and occur mainly on the near side.
- **MARE** Dark lowland plains of hardened basalt rocks
- **RILL** Ancient lava channels.. They include Hyginus Rill and Hadley Rill.
- **NAMES** Most lunar features are named after famous scientists, especially astronomers.

NEAR AND FAR

- The Moon turns around once in the same time it takes to go around the Earth once.
- This means that it keeps mainly one side facing Earth.
- Due to the Moon's slight variations in orbit and the 'wobble' on its axis, a total of almost three-fifths of its surface is visible from Earth.
- The other two-fifths is always hidden and has only been seen by spacecraft in lunar orbit.
- The far side is sometimes called the 'dark side of the Moon' but it receives sunlight in the same pattern as the near side.

PHASES OF THE MOON

The Sun lights up only half the Moon at a time. As the Moon moves around the Earth, we see varying amounts of the sunlit half of the Moon. This causes it to show changes of shape called phases. When the Moon is roughly between the Sun and the Earth, its dark side is turned towards us, and we normally cannot see it. This is called the New Moon. As the Moon moves around the Earth, the sunlit side begins to show. First we see a thin crescent, then a half Moon, and then a Full Moon. At Full Moon, all of the sunlit side faces us. After Full Moon the phase slowly decreases to half and back to a crescent as we see less and less of the sunlit side. Finally it is New Moon once again. The time from one New Moon to the next New Moon is 29.53 days.

THE MOON AND TIDES

Tides are the regular rising and falling of the surface of the oceans. Although the Sun has some influence, ocean tides are mainly caused by the gravitational interaction between the Earth and Moon. The gravitational pull of the Moon causes the oceans to bulge out in the direction of the Moon. Another bulge occurs on the opposite side, since the Earth is also being pulled towards the Moon, and away from the water on the far side. Since the Earth is spinning on its axis, tides rise and fall twice a day, the interval between low tide and high tide being just over six hours.

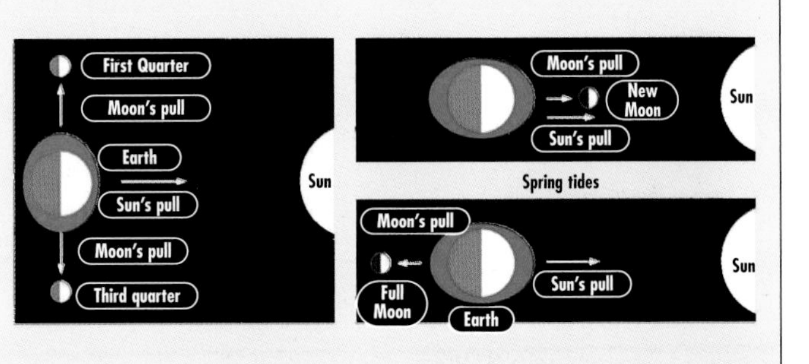

ECLIPSES

Solar eclipse
When the Moon comes between the Earth and Sun and blocks out part (partial eclipse) or all (total eclipse) of the Sun's disc. The amount of the disc blotted out varies with the position on Earth. The area of shadow on Earth of a total eclipse is 270 km wide and moves across the Earth as the Earth spins and the Moon continues its orbit. On average there are 2 total eclipses every 3 years.

Lunar eclipse
When the Earth comes between the Sun and Moon. The Moon seems to fade but stays shining a copper-red colour, due to sunlight rays bent (refracted) around the edge of the Earth by the atmosphere.

THE MOON TIMELINE

1959
Russian space probe Luna 2 crashed into the surface in September - the first human-made object to reach another world. Next month Luna 3 went around the Moon and sent back the first images of the previously unknown far side.

1966
Luna 9 soft-landed and sent back the first close-up images of the Moon's surface. Lunar 10 became the first probe to go into steady Moon orbit.

1966
The first of the US Surveyor missions touched down in June and sent back more than 11,000 images.

1968
In December US Apollo 8 went into Moon orbit but did not land, and the crew stayed on board the command module in lunar orbit.

1969
US Apollo 11 touched down on 20 July, carrying the first humans to visit another world. Neil Armstrong was first to step out of the Lunar Module (lander) onto the surface, followed by Edwin 'Buzz' Aldrin. Michael Collins stayed on board the command module in lunar orbit.

1969
Apollo 12 landed to carry out scientific studies in November.

1971
Apollo 14 collected 43 kg of Moon rocks in February.

1972
Apollo 16 collected almost 100 kg of Moon material.

1972
Eugene Cernan was the last person to step off the Moon.

1994
The Clementine space probe collects information that suggestions there might be frozen water on the Moon.

2003
European spacecraft Smart 1 launched.

2004
In February US President George W Bush announced plans for a new series of Moon missions.

2004
Smart 1 took up lunar orbit on 15 November, to study and map the surface using X-rays.

MARS

Named after the Roman god of war, Mars is also called the 'Red Planet' because its surface rocks and dust contain large amounts of the substance iron oxide – better known here on Earth as rust. Like Earth, Mars has polar ice caps, volcanoes, canyons, winds and swirling dust storms. Features resembling river beds, estuaries and shorelines suggest that great rivers – probably of water – once flowed across Mars' surface. Despite many visits by space probes, landers and rovers, no signs of life have ever been found.

WHERE IN THE SOLAR SYSTEM?

SURFACE CONDITIONS

ATMOSPHERE: Mostly carbon dioxide, small amounts of nitrogen and argon, traces of oxygen, carbon monoxide and water vapour.

NATURE OF SURFACE: Rocks and dust, including giant volcanoes, deep canyons and dusty plains.

AVERAGE SURFACE TEMPERATURE: minus 63°C

LOWEST SURFACE TEMPERATURE: minus 140°C

HIGHEST SURFACE TEMPERATURE: 20°C

WEATHER OR CLIMATE: Clouds, fog, strong winds, dust storms, red sky.

SEASONAL CHANGES: Marked (similar to Earth) with intensely cold winters.

A photograph of the surface of Mars taken by the Viking lander.

PLANET PROFILE

Diameter at Equator **6,804 km**

Surface area **144 million sq km**

Tilt of axis **25.1°**

Mass (Earth = 1) **0.107**

Volume (Earth = 1) **0.151**

Overall density **3.9 g per cc**

Gravity (Earth = 1) **0.38**

Number of moons **2**

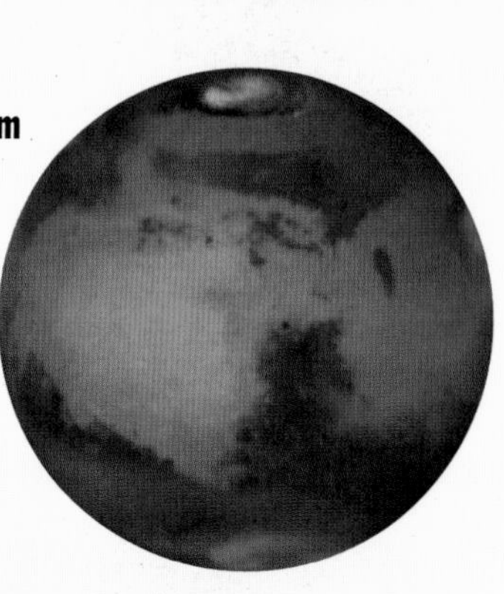

A NASA photograph of Mars.

ORBIT DETAILS

Average distance from Sun 227.9 million km

Average distance from Sun 1.52 AU (Earth = 1)

Closest distance to Sun (perihelion) 206.6 million km

Farthest distance from Sun (aphelion) 249.2 million km

Average orbital speed 24.1 m per second

Slowest orbital speed 21.9 m per second

Fastest orbital speed 26.5 m per second

Time for one orbit (Mars year) 686.9 Earth days

Axial rotation period (Mars day) 24.62 Earth hours

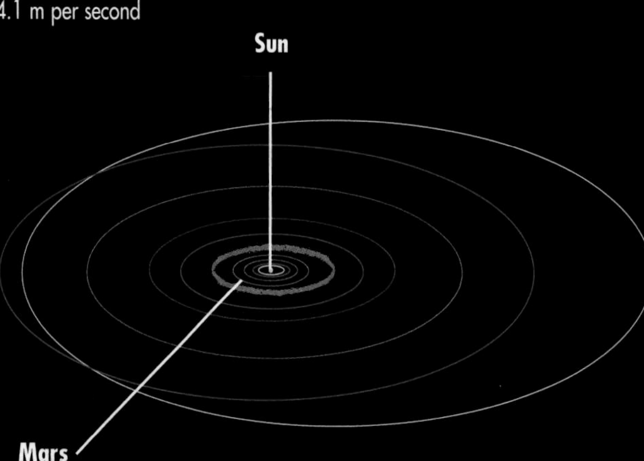

MAJOR FEATURES

North Polar Cap

- Water ice that remains through summer.
- Sand dunes formed by wind.
- North polar cap is approx. 680 miles across.

South Polar Cap

- The frost contains frozen carbon dioxide.
- Carbon dioxide freezes at around –125°C.
- South polar cap is approx. 260 miles across.

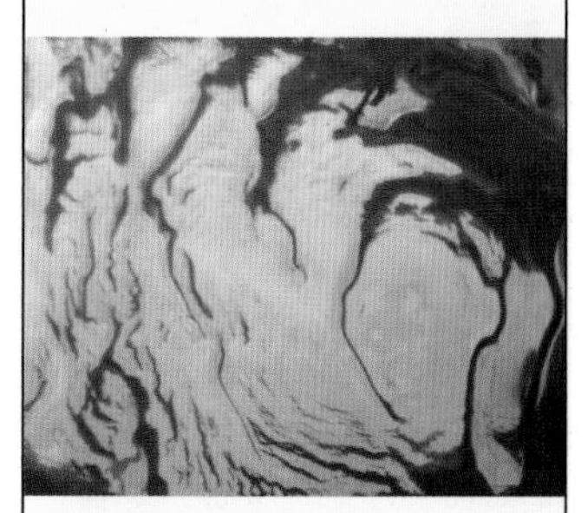

Southern Polar Cap

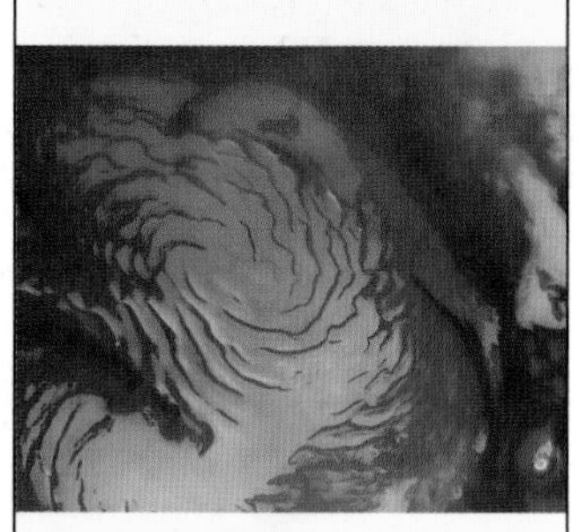

Northern Polar Cap

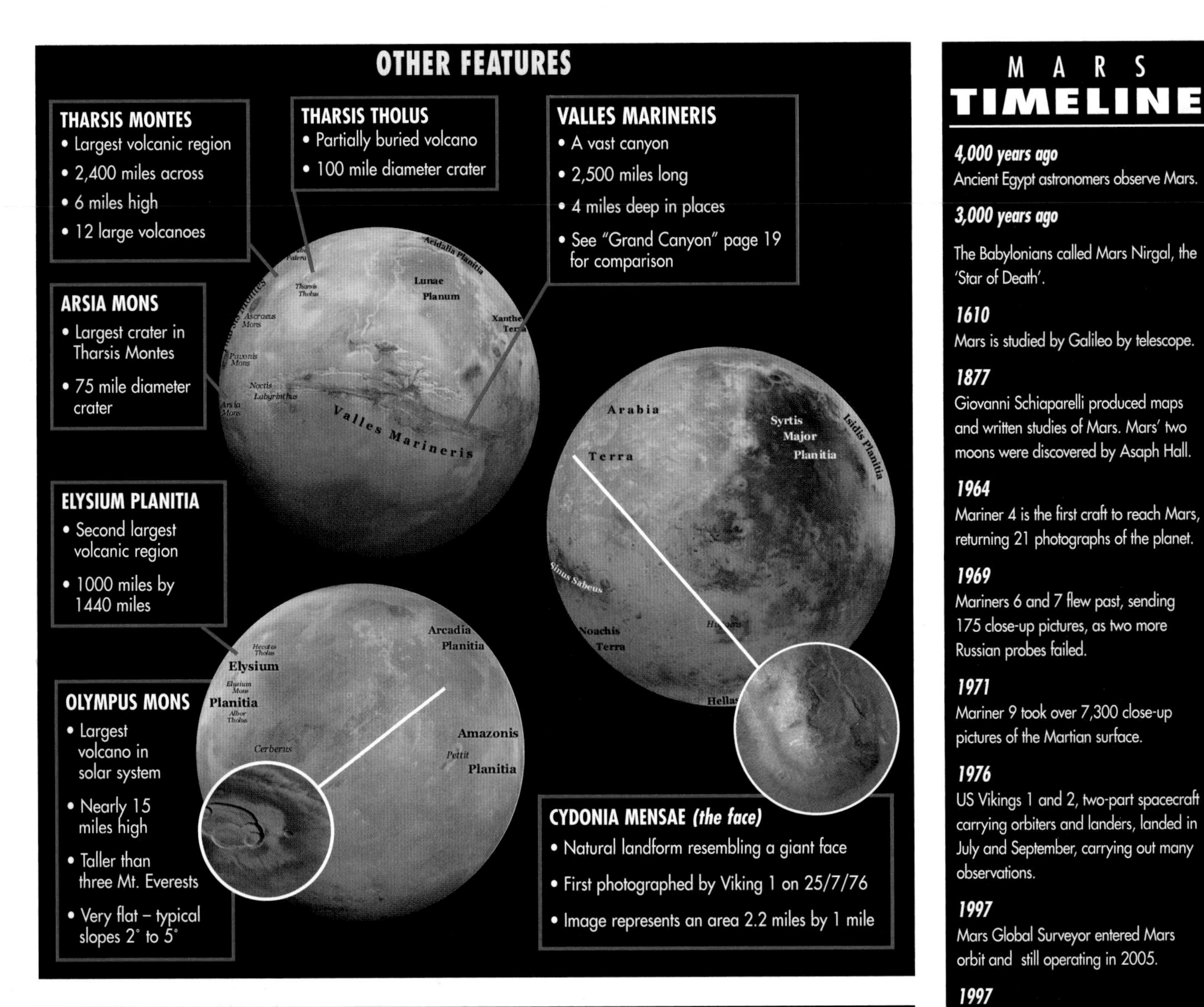

MARTIANS!

- Giovanni Schiaparelli's studies of Mars in 1877 used the term 'canali' meaning channels, which could be naturally occurring.
- Some people took this to mean 'canals' as made by some advanced life-form, like we dig on Earth.
- Percival Lowell developed the idea to suggest Martians dug canals to take water from the planet's ice-caps to water their crops, since other areas on Mars changed colour with the seasons.
- The myth of Martians began, and H G Wells featured their invasion of Earth in *War of the Worlds* (1898), which continues to be popular and made into films.
- The channels are now known to be imagined or perhaps long-dry watercourses, and the colour changes are probably dust storms.

• See pages 55 for information on space probes to Mars.

MARS TIMELINE

4,000 years ago
Ancient Egypt astronomers observe Mars.

3,000 years ago
The Babylonians called Mars Nirgal, the 'Star of Death'.

1610
Mars is studied by Galileo by telescope.

1877
Giovanni Schiaparelli produced maps and written studies of Mars. Mars' two moons were discovered by Asaph Hall.

1964
Mariner 4 is the first craft to reach Mars, returning 21 photographs of the planet.

1969
Mariners 6 and 7 flew past, sending 175 close-up pictures, as two more Russian probes failed.

1971
Mariner 9 took over 7,300 close-up pictures of the Martian surface.

1976
US Vikings 1 and 2, two-part spacecraft carrying orbiters and landers, landed in July and September, carrying out many observations.

1997
Mars Global Surveyor entered Mars orbit and still operating in 2005.

1997
Mars Pathfinder lands and its rover Sojourner explored the Martian surface.

1999
The US Mars Climate Orbiter and Mars Polar Lander reached the planet but both fell silent.

2001
US Mars Odyssey successfully reached Mars orbit and has sent back a wealth of scientific information.

2003
European Mars Express entered orbit but its Beagle 2 lander was lost.

2004
Mars Exploration rovers Spirit and Opportunity touched down and began to explore the Martian surface and send back a wealth of information.

2005
Spirit and Opportunity continue to rove and send back data.

2025
Proposed end date for possible US missions carrying astronauts to Mars.

JUPITER

By far the biggest planet in the Solar System, Jupiter is a vast world of swirling gases and storms of unimaginable fury. As the fifth planet out, it is the nearest 'gas giant' to the Sun. It is not much smaller than some of the stars called brown dwarfs, but it does not shine itself, and reflects sunlight as all planets do. Even so, its huge pull of gravity holds more than 60 moons in orbit around it. Jupiter is named after the Roman king of the gods, also called Jove.

ATMOSPHERIC CONDITIONS

ATMOSPHERE:
Mostly hydrogen, hydrogen sulphide, some helium, traces of methane, water vapour, ammonia and other gases.

NATURE OF SURFACE:
Visible surface is whirling gases, possibly a solid surface on a small rocky core km below visible surface.

AVERAGE TEMPERATURE:
minus 130°C

LOWEST TEMPERATURE:
minus 163°C

HIGHEST TEMPERATURE:
minus 170°C

WEATHER OR CLIMATE:
Complete cloud coverage with storms and wind speeds up to 600 km/h.

SEASONAL CHANGES:
Few, being so far from Sun.

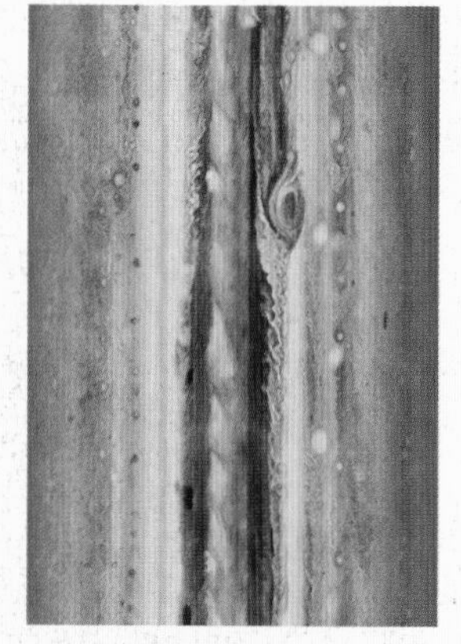

A diffuse shot of Jupiter's surface.

PLANET PROFILE

Diameter at Equator **142,984 km**

Surface area **61.4 billion sq km**

Tilt of axis **3.13°**

Mass (Earth = 1) **318**

Volume (Earth = 1) **1,236**

Overall density **1.33 g per cc**

Gravity (Earth = 1) **2.36**

Number of moons **60-plus and counting**

A NASA photograph of Jupiter.

ORBIT DETAILS

Average distance from Sun
778.4 million km

Average distance from Sun
5.203 AU (Earth = 1)

Closest distance to Sun (perihelion) 740.7 million km

Farthest distance from Sun (aphelion) 816.1 million km

Average orbital speed
13.0 km per second

Slowest orbital speed
12.4 km per second

Fastest orbital speed
13.7 km per second

Time for one orbit
(Jupiter year) 11.87 Earth days

Axial rotation period
(Jupiter day) 9.92 Earth days

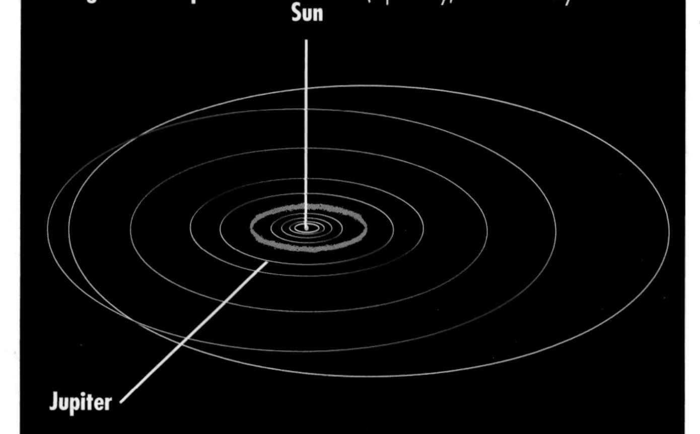

MAJOR FEATURES

Jupiter has several distinctive features mapped by astronomers.

Great Red Spot
A giant storm system three times wider than Earth, that travels around Jupiter just south of the equator, once every 6 days.

White Spots
Smaller circulatory storm system in Jupiter's atmosphere, siimilar in size to the Earth.

Brown Spots
Stormy regions that are probably warmer than surrounding clouds.

Rings
These consist of dust knocked from Jupiter's moons by meteor strikes.

Inner Structure
Central small rocky core; then a layer of 'metallic' hydrogen; then liquid hydrogen; and finally the outermost atmosphere of mainly hydrogen gas. These layers grade from one to another, with no sharp boundaries.

Jupiter's Giant Red Spot.

• See page 55 for information on probes to Jupiter.

OTHER FEATURES

- **BELTS** Strips of dark clouds which snake along from west to east (left to right) and change through the years.
- **ZONES** Lengths of light-coloured clouds which change like the darker belts. Zones contain higher clouds than belts. Blue-tinted clouds are lowest and warmest.
- **TURBULENCE** Belts sometimes move in the opposite direction to their neighbouring zones, creating swirling patterns of storms and turbulence along their edges.

MOON RECORDS

- Ganymede is the largest moon in the Solar System.
- Callisto is the most heavily cratered object.
- Io probably has the most volcanic activity.

A NASA collage showing IO's red and black coloured lava flows and yellow sulphur patches.

JUPITER'S RINGS

- **HALO RING** Faint inner ring 10,000 km thick.
- **MAIN RING** Dust from Adrastea and Metis, 100 km thick.
- **INNER GOSSAMER RING** Dust from Amalthea, 2,600 km thick.
- **OUTER GOSSAMER RING** Dust from Thebe 8,800 km thick
- **FAINT OUTER RING** Rings beyond outer Gossamer Ring, 8,800 km thick

TRUE GIANT

Jupiter has more than twice as much mass ('weight') than all the other eight planets added together. However it would probably need to be 50 times heavier to start burning like a true star.

SPEED SPIN

Jupiter is not only the largest planet, it also spins around the fastest, once in less than 10 Earth hours. The spinning speed of the upper atmosphere at the equator is 5 minutes faster than at the poles, so the atmosphere is continually being twisted and torn.

JUPITER'S MOONS

On 7-11 January 1610 Galileo discovered Jupiter's four main moons, now known as Galilean moons, following their orbits across the face of the planet. This was direct evidence that the Earth was not at the centre of everything, and strengthened his idea that planets like Earth and Jupiter probably revolved around the Sun.

MOON (or group)	DIAMETER	DISTANCE FROM JUPITER
Inner group (Metis, Adrastea, Amalthea, Thebe)	Four small moons less than 200 km across	Less than 200,000 km
Io	3,642 km	421,700 km
Europa	3,122 km	671,030 km
Ganymede	5,262 km	1,070,000 km
Callisto	4,820 km	1,882,000 km
Themisto	8 km	7,392,000 km
Himalia group	Most under 100 km	11-12 million km
Ananke group	Most under 30 km	21 million km
Carme group	Most under 5 km	23-24 million km
Pasiphaë	Small outermost moons	23 million km

JUPITER TIMELINE

3,000 years ago
Jupiter was known to the Greeks and then the Romans.

1,500 years ago
In Ancient China Jupiter was known as the 'Wood Star'.

1610
Galileo observed Jupiter's four largest moons.

1665
The Great Red Spot was first observed.

1690
Giovanni Domenico Cassini noticed that the upper atmosphere takes longer to spin at the poles than around the equator.

1973
US Pioneer 10 probe flew past Jupiter.

1979
Voyager 1 flew past taking spectacular photographs. Voyager 2 achieved similar results.

1992
Ulysses probe passes by Jupiter on its way to the Sun taking measurements.

1994
Parts of comet Shoemaker-Levy 9 hit Jupiter in July, photographed by the approaching Galileo space probe.

1995
Galileo became the first probe to go into orbit around Jupiter, starting on 7 December. On the same day an atmosphere probe it had already released parachuted 150 km into the atmosphere, collecting information.

1996-2003
Galileo continued its studies of Jupiter and its nearer moons.

2000
The Great Red Spot had shrunk to about half the size it measured in 1900.

2000
Cassini probe passes on way to Saturn.

2003
Galileo finally plunged into the clouds in September.

2007
The New Horizons probe is due to fly past, on its way to Pluto.

2010
The US Jupiter probe Juno is scheduled for launch. Juno will orbit over Jupiter's poles.

SATURN

Long famed for its glistening, breathtakingly beautiful rings, Saturn is the Solar System's second-largest planet after its neighbour Jupiter. Saturn was the Roman god of farming, civilization and prosperity – and also the name of the rockets that powered Apollo astronauts to the Moon. Due to its fast spin, 'gas giant' make-up and very light weight compared to its size, Saturn bulges around its middle, or equator, as it rotates. This means the planet is fatter than it is tall by almost 12,000 kilometres.

WHERE IN THE SOLAR SYSTEM?

ATMOSPHERIC CONDITIONS

ATMOSPHERE:
Mostly hydrogen, small amount of helium, traces of methane, water vapour, ammonia

NATURE OF SURFACE:
Visible surface is whirling gases, possibly a solid surface on a rocky core km below visible surface

AVERAGE TEMPERATURE:
minus 130°C

LOWEST TEMPERATURE:
minus 190°C

HIGHEST TEMPERATURE:
minus 120°C

WEATHER OR CLIMATE:
Clouds and storms of fast-moving gases, high wind speeds

SEASONAL CHANGES:
Few, being so far from Sun

Saturn's northern hemisphere is presently a serene blue, much like that of Uranus or Neptune.

PLANET PROFILE

Diameter at Equator **120,536 km**

Surface area **42.7 billion sq km**

Tilt of axis **26.7°**

Mass (Earth = 1) **95.2**

Volume (Earth = 1) **688.9**

Overall density **0.69 g per cc**

Gravity (Earth = 1) **0.91**

Number of moons **50-plus**

A NASA photograph of Saturn.

ORBIT DETAILS

Average distance from Sun
1,426 million km

Average distance from Sun
9.54 AU (Earth = 1)

Closest distance to Sun (perihelion) 1,349 million km

Farthest distance from Sun (aphelion) 1,504 million km

Average orbital speed
9.6 km per second

Slowest orbital speed
9.1 km per second

Fastest orbital speed
10.1 km per second

Time for one orbit
(Saturn year) 29.46 Earth days

Axial rotation period
(Saturn day) 10.77 Earth hours

Sun

Saturn

MAJOR FEATURES

Saturn has been explored by a number or probes.

Rings
These are made of billions of fragments of ice and rock which reflect sunlight, glisten and sparkle. The largest particles are car-sized.

- Fainter, more distant rings
- Outermost main ring A
- Middle ring B
- Innermost main ring C
- Innermost ring D
- Cassini Division
- Encke Division

Equatorial Zone
Rotates about 25 minutes faster per Saturn day than the Polar Zones, with wider cloud banding.

North and South temporate zones Clouds and winds of 1,800 km/h.

South Polar Zone
Lighter coloured clouds, and a warm dark spot.

Saturn's poles are shown in this false-colour NASA image.

OTHER FEATURES

- **CORE** Saturn's core is probably very hot, nearly 12,000°C, and the planet gives out more heat energy than it receives from the Sun.
- **SOUTH POLE** A very hot region that glows bright on infra-red (heat-ray) photographs.
- **BAND CLOUDS** Less marked than on Jupiter, consisting of stripes and zones of clouds at different temperatures. They tend to be wider nearer the equator.
- **WHITE SPOTS** Tend to come and go, probably areas of swirling gases.

MAIN MOONS

After the discovery of Titan by Huygens, the second to fifth moons were spotted by Giovanni Domenico Cassini.

YEAR	MOON	DIAMETER	DISTANCE
1684	Tethys	1,060 km	294,600 km
1684	Dione	1,120 km	377,400 km
1672	Rhea	1,530 km	527,100 km
1671	Iapetus	1,435 km	3,560,000 km

TITAN

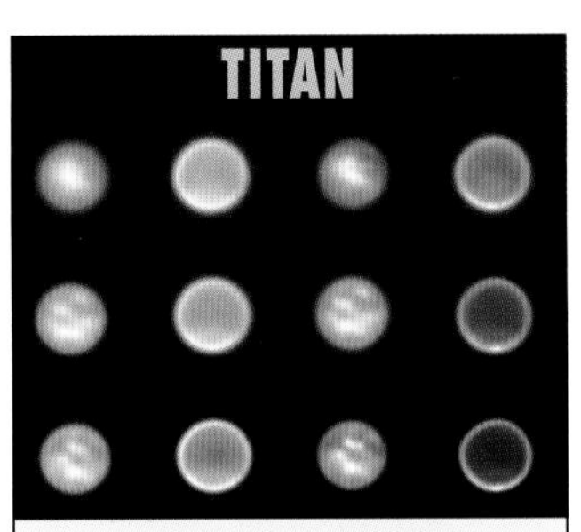

A selection of shots of Titan from the European Southern Observatory.

- Titan is the second-largest moon in the Solar System, after Jupiter's Ganymede.
- It is the moon with a thick atmosphere.
- Its atmosphere is mainly nitrogen (like Earth's), plus methane, ethane, acetylene, propane, carbon dioxide, carbon monoxide, hydrogen cyanide and helium.

VAST BUT LIGHT

Saturn is the only planet whose density, or mass per volume, is less than water. We would say it is 'lighter than water' and if there was a tank of water big enough to run a bath for this enormous planet, Saturn would float in it!

THE RINGS OF SATURN

The rings of Saturn were first noticed by Galileo, who could not quite make them out with his early telescope. He guessed they might be moons, one on each side, and called them 'ears of Saturn'.
They appear to change in shape when viewed from Earth, as they are tilted and slowly turn with Saturn's orbit, so we see them at different angles. Viewed edge-on, they are at their thinnest, about every 15 years. Each main ring is made of thousands of smaller 'ringlets'.

Name	Inner edge distance from Saturn km	Width km
D Ring	6,000	7,150
C Ring	14,500	17,500
Columbo Gap	17,800	100
Maxwell Gap	27,500	270
B Ring	32,000	25,500
Cassini Division	57,500	4,700
A Ring	62,200	14,600
Encke Division	73,570	325
F Ring	80,200	500
G Ring	104,000	8,000
E Ring	120,000	300,000

• *See page 10 for information on Galileo.*

SATURN TIMELINE

2,500 years ago
Saturn was known to Greeks and then the Romans.

1610
Galileo saw two shapes on either side of Saturn, the first observations of its rings.

1655
Christiaan Huygens discovered Titan and gives true explanation of Saturn's ring system.

1789
William Herschel discovers that Saturn bulges at the equator and is flattered at the poled.

1847
John Herschel (son of William Herschel) named the then-known 7 moons of Saturn.

1979
Pioneer was the first space probe to visit Saturn.

1980
Voyager 1 sent back the first clear pictures of the planet.

1981
Voyager 2 flew past in August and discovered further features including darker 'spoke' regions in Saturn's B ring, smaller gaps between rings and more moons.

1997
Cassini-Huygens space probe launched on 15 October.

2004
Cassini-Huygens flew close to moon Phoebe in June and reached Saturn orbit on 1 July.

2004
After two Titan flypasts, the Huygens lander was released from the Cassini orbiter on 25 December.

2005
Huygens plunged into Titan's atmosphere on 14 January, sending information after touching down. Cassini continued to orbit and fly past many moons, especially Titan.

2008
Expected end of main mission for Cassini orbiter, but this may be extended.

URANUS

Uranus is the third 'gas giant' and seventh planet out from the Sun. It is very similar in size and structure to its next-neighbour-out, Neptune, being partly gas but also containing much rocky and frozen material. The axis of Uranus is almost at right angles to the Sun, so Uranus seems to 'lie on its size and roll like a blue marble around the Sun'. The planet is named after the Greek ruler of the Heavens, and the father of Saturn.

WHERE IN THE SOLAR SYSTEM?

ATMOSPHERIC CONDITIONS

ATMOSPHERE:
Mostly hydrogen, about one-sixth helium, also methane, traces of ammonia.

NATURE OF SURFACE:
Gassy, with any solid surface deep below; glows in sunlight as bright blue-green or cyan.

AVERAGE TEMPERATURE:
minus 205°C

LOWEST TEMPERATURE:
minus 214°C

HIGHEST TEMPERATURE:
minus 200°C

WEATHER OR CLIMATE:
Swirling clouds, winds and gases, despite smooth 'glassy' appearance.

SEASONAL CHANGES:
Extreme, since Uranus lies on its side so that during each orbit both poles and the equatorial regions face the Sun.

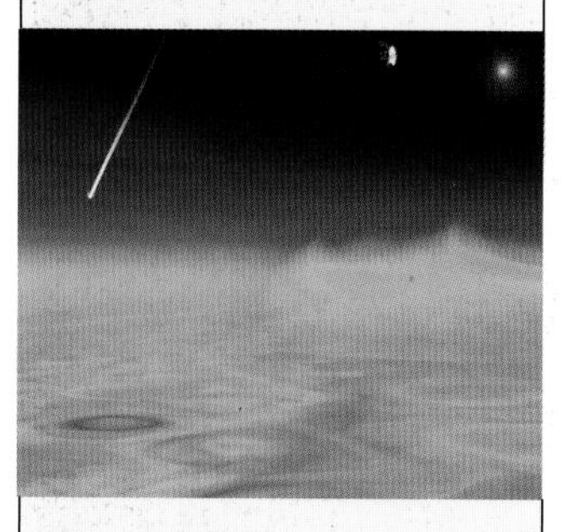

This rendered image shows the Voyager spacecraft in the sky above Uranus.

PLANET PROFILE

Diameter at Equator **51,118 km**

Surface area **8.08 billion sq km**

Tilt of axis **97.8° (almost at right angles to Sun)**

Mass (Earth = 1) **14.54**

Volume (Earth = 1) **63.1**

Overall density **1.32 g per cc**

Gravity (Earth = 1) **0.89**

Number of moons **approaching 30 and counting**

A NASA photograph of Uranus.

ORBIT DETAILS

Average distance from Sun
2,870 million km

Average distance from Sun
19.2 AU (Earth = 1)

Closest distance to Sun (perihelion) 2,736 million km

Farthest distance from Sun (aphelion) 3,006 million km

Average orbital speed
6.79 km per second

Slowest orbital speed
6.68 km per second

Fastest orbital speed
7.13 km per second

Time for one orbit
(Uranus year) 84.1 Earth years

Axial rotation period
(Uranus day) 17.24 Earth hours

Sun

Uranus

MAJOR FEATURES

Uranus has been mapped from Earth and by Voyager 2.

Inner Structure
Uranus is probably quite similar in composition all the way through, with gases and particles of rocks and ice intermingled.

Colour
Uranus usually appears pale to mid blue-green, a colour known as cyan, probably because methane crystals in its atmosphere absorb most of the red light in sunlight, leaving it mainly blue.

Streaking
Hubble Space telescope images reveal faint streaks which slowly change, perhaps due to seasonal variations.

Warmer Equator
Despite Uranus's extreme tilt, the equator is slightly warmer than the polar regions.

Artist's impression of the rings of Uranus.

OTHER FEATURES

- **SURFACE FEATURES** There are few obvious features when Uranus is viewed through telescopes from Earth, the surface being smooth with a 'satin' glow, almost like polished marble.
- **MAGNETIC FIELD** This invisible force field is 'offset', with its centre not in the centre of the planet (like Neptune), and it is also tilted at 60° compared to the planet's spinning axis.

ON ITS SIDE

The axis of Uranus is almost at right angles to the Sun and the other almost 'upright' planets.So Uranus spins as if lying on its side, rolling around the Sun.- The axis of Uranus does not swing around as it orbits the Sun, as if a long piece of string joined it to the Sun, but points the same way.- So the South Pole of Uranus (pointing sideways) faces the Sun for a short time. Then as the orbit continues, the North Pole gradually comes around the face the Sun, in the opposite part of the orbit.

MAIN MOONS

MOON	DIAMETER	DISTANCE
Miranda	472 km	129,870 km
Ariel	1,158 km	190,910 km
Umbriel	1,170 km	266,000 km
Titania	1,578 km	436,300 km
Oberon	1,522 km	583, 520 km

There are about 14 smaller moons inside the orbit of Miranda, most discovered by Voyager 2. The biggest is Puck, measuring about 160 km across.

RINGS OF URANUS

Ring	Distance from centre of Uranus (km)	Width of ring (km)
Lambda	38,000	2,500
6	41,840	1–3
5	42,230	2–3
4	42,580	2–3
Alpha	44,720	4-10
Beta	45,670	5-11
Eta	47,190	0–2
Gamma	47,630	1–4
Delta	48,290	3-7
1986U1R	50,020	1–2
Epsilon	51,140	20–100

ODD LITTLE WORLD

- Uranus's principal innermost moon, Miranda, has one of the oddest appearances in the Solar System.
- Massive canyons scar the surface, as well as mountains, cliffs and craters.
- Three huge race-track-like shapes called ovoids are prominent, perhaps formed by upwelling of rocks from within.
- Miranda's Verona Rupes is a huge fault scarp. At 20 km high it is the tallest cliff in the Solar System
- Miranda may have frozen water, methane-type substances and rocks on its surface.

A composite image of Uranus's strange moon Miranda.

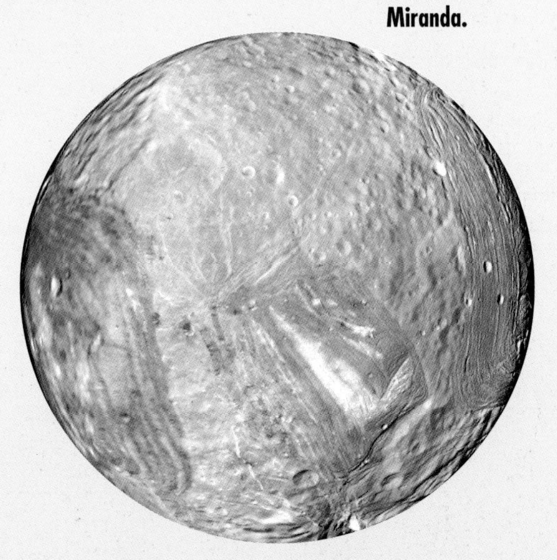

• *See pages 18-19 for information on Earth.*

URANUS TIMELINE

Ancient times
Uranus was not known as a planet to ancient people.

1690
John Flamsteed recorded Uranus as a dim star-like object, 34 Tauri.

1748
James Bradley observed Uranus as a faint star, and did the same in 1750 and 1753.

1764
Pierre Charles Le Monnier recorded Uranus a dozen times from this year to 1771.

1781
William Herschel discovers Uranus.

1787
William Herschel discovered moons Titania and Oberon.

1851
William Lassell discovered Ariel and Umbriel, Uranus's second- and third-nearest 'twin moons', on 24 October.

1948
Gerard Kuiper discovered Miranda, Uranus's innermost moon, on 16 February.

1977
Voyager 2 was lauched on its great journey across the Solar System.

1977
In March 1977 a system of rings was suspected, as the planet blotted out a faint star behind it in an odd manner.

1982
Voyager 2 passed Uranus, at the time when its South Pole pointed directly towards the Sun.

1986
Voyager 2 made its closest flypast during January, some 110,000 km from its surface. It saw the rings in detail and discovered an extra one, also 10 more moons in addition to the five visible from Earth with telescopes.

2007
The Sun will be overhead at the equator of Uranus, midway between its apparent journey from being directly over one pole to overhead at the other pole.

NEPTUNE

WHERE IN THE SOLAR SYSTEM?

The deep blue colour of this fourth 'gas giant' inspired its name of Neptune, Roman god of the sea. Neptune's atmosphere is in fact ravaged by the fastest winds in the Solar System. Although it is the fourth largest planet, it is third heaviest, being denser than its similar-sized neighbour, Uranus. Also like Uranus, Neptune's atmosphere probably extends about one-fifth of the way towards to the centre. Then it gives way to a mix of semi-liquid ice, rocks, methane and ammonia, with a central core of chiefly part-molten rocks and metals.

ATMOSPHERIC CONDITIONS

ATMOSPHERE:
Mostly hydrogen, one-fifth helium, traces of methane, ethane.

NATURE OF SURFACE:
Gassy, with incredibly fast wind movements.

AVERAGE TEMPERATURE:
minus 220°C

LOWEST TEMPERATURE:
minus 223°C

HIGHEST TEMPERATURE:
minus 218°C

WEATHER OR CLIMATE:
Storms and swirling gases move at 2,000 km/h, more than four times quicker than Earth's fastest winds in tornadoes (twisters).

SEASONAL CHANGES:
Neptune has few seasonal changes. Since it is so far from the Sun, solar heat and light have little effect on this icy cold planet.

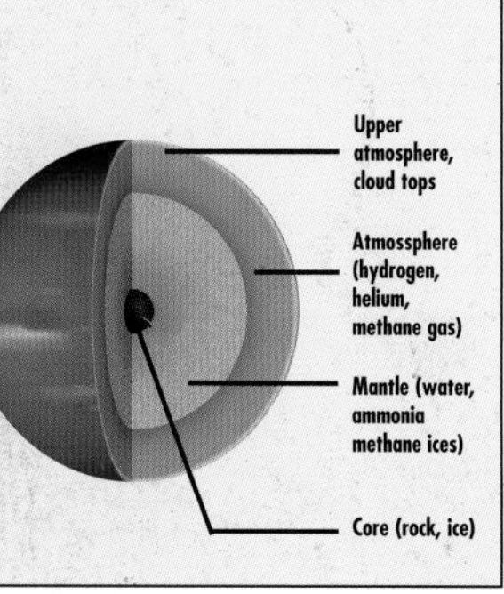

PLANET PROFILE

Diameter at Equator **49,528 km**

Surface area **7.62 billion sq km**

Tilt of axis **28.3°**

Mass (Earth = 1) **17.15**

Volume (Earth = 1) **57.7**

Overall density **1.64 g per cc**

Gravity (Earth = 1) **1.14**

Number of moons **about 13**

A NASA photograph of Uranus. The great dark spot is in the centre.

ORBIT DETAILS

Average distance from Sun
4,498 million km

Average distance from Sun
30.1 AU (Earth = 1)

Closest distance to Sun (perihelion) 4,459 million km

Farthest distance from Sun (aphelion) 4,536 million km

Average orbital speed
5.43 km per second

Slowest orbital speed
5.39 km per second

Fastest orbital speed
5.48 km per second

Time for one orbit
(Neptune year) 164.8 Earth years

Axial rotation period
(Neptune day) 16.1 Earth hours

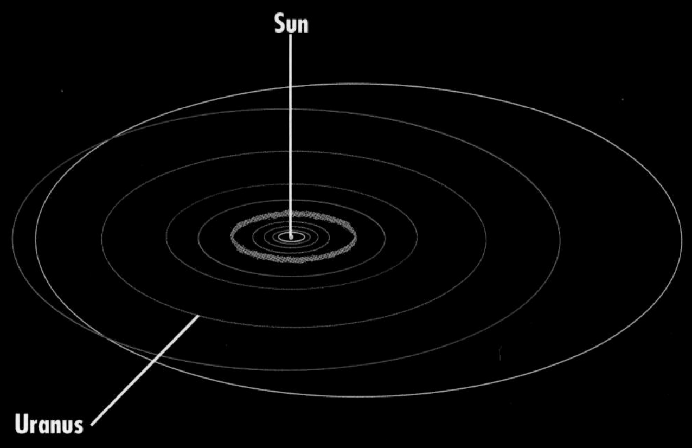

MAJOR FEATURES

Neptune has been mapped from Earth and by probes.

Great Dark Spot
About as wide is Earth, this was probably a vast storm system of swirling gases. It faded in the mid 1990s but another similar area, GDS2, appeared.

Colour
Neptune is a dark blue-green, probably because methane crystals in its atmosphere absorb most of the red light in sunlight, leaving it mainly blue.

Winds
Neptune has some of the fastest winds in the Solar System, blasting along at over 2,300 km/h .

Inner Structure
Outer atmosphere of hydrogen and helium, then lower down more methane and ammonia, with more rock particles, merging into melted rock and metal in the central core.

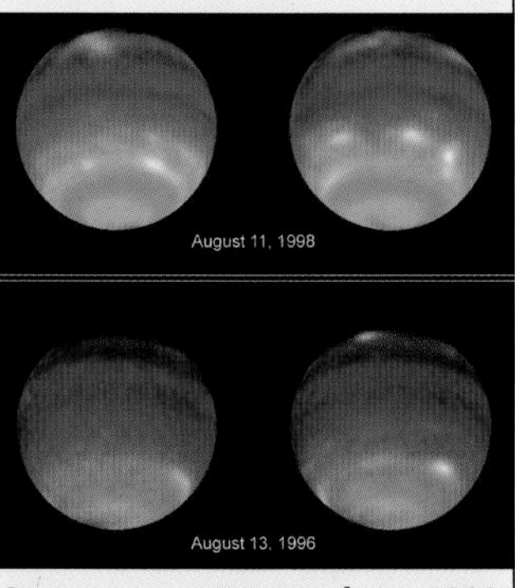

Storms rage on Neptune during 1998.

OTHER FEATURES

- **MAGNETIC FIELD** Tilted at 47° to the axis around which Neptune spins, and also off centre from the middle of the planet by more than 13,000 km.
- **OUTER CLOUDS** Some clouds high above the main cloud layer cast shadows on those below, much as Earth's clouds cause shadows on the land.
- **RINGS** About 9 faint rings surround Neptune, with a strange structure showing clumps and lumps of larger material rather than spread-out small particles.
- **ARCS** Curved arcs within the outermost ring, Adams, are probably due to the movements and gravity of the moon Galatea on their inner side.

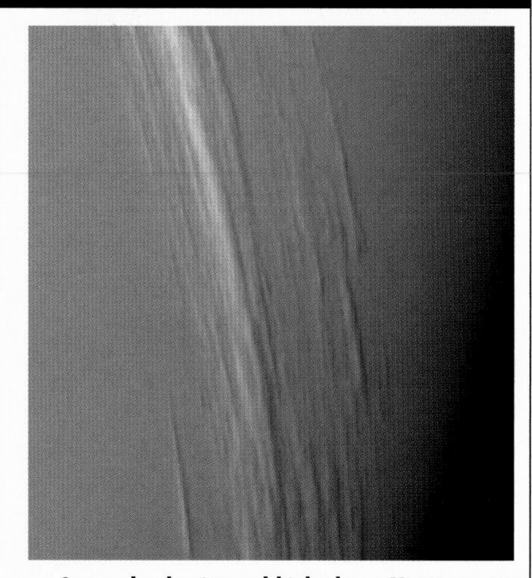

Outer clouds pictured high above Neptune.

THE FIRST PAPER PLANET

Neptune was the first planet to be 'discovered' on paper, from calculations of the orbit of Uranus, which showed another body beyond it affects its movements (see Timeline, 1846).

NAMING NEPTUNE

- Soon after Neptune was discovered, several astronomers suggested names, often inspired by its deep blue-green sea-like colour:
- Various experts including Arago proposed Leverrier after its co-discoverer.
- Others introduced Poseidon (right), Greek sea god.
- Galle proposed Janus.
- Challis suggested Oceanus.
- Le Verrier himself proposed Neptune.
- The name Neptune was adopted by the end of 1946.

NEPTUNE'S RINGS

These are very faint, difficult to measure and also seem to change rapidly.

Name	Distance from surface (km)	Width (km)
Galle (1989 N3R)	17,100	15
Leverrier (1989 N2R)	28,450	15
Lassell (1989 N4R)	30,650	6
Arago	32,850	not clear
Adams (1989 N1R)	38,150	less than 50
Liberty Arc	leading arc	not known
Equality Arc	middle arc	not known
Fraternity Arc	trailing arc	not known
Courage Arc	not known	not known

VOYAGER VISIT

- Voyager 2 is the only probe to visit Neptune, and got closest to this planet than to any other on its journey from Earth.
- Period of observation June to October 1989.
- At its closest on 25 August, the probe passed just 5,000 km above Neptune's North Pole.
- A few hours later it passed within 40,000 km of the largest moon, Triton.
- Triton was Voyager 2's last studied object before it left the Solar System.

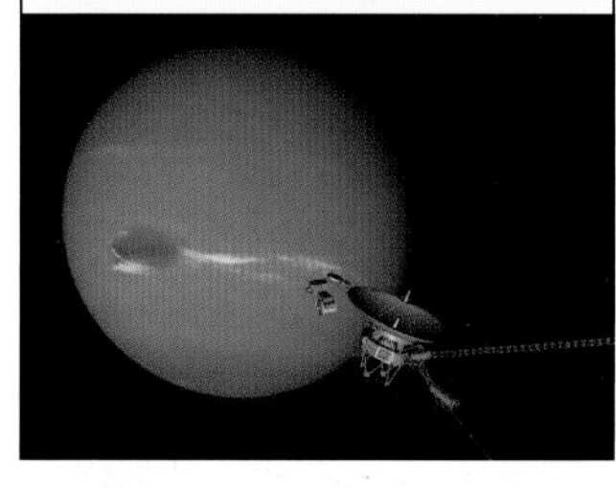

SIZE AND SHAPE

Neptune is the fourth biggest planet in size, being slightly wider than its next-neighbour-in Uranus. But because Neptune is size-for weight heavier, or more dense, than Uranus, it is the third heaviest planet after Jupiter and Saturn.

NEPTUNE TIMELINE

1612
Galileo saw Neptune on 28 December but didn't recognise it as a planet.

1843
John Adams calculated that probably another planet farther out than Uranus was affecting its orbit.

1846
In August, the British astronomer James Challis twice observes Uranus using Adams' predictions, but does not recognize the planet.

Neptune was officially discovered by Johann Galle at the Berlin Observatory on 23 September . He used calculations worked out by Urbain le Verrier.

1846
Neptune's largest moon Triton identified by William Lassell.

1880
Camille Flammarion proposed the name Triton for Neptune's main moon.

1949
Neptune's second moon Nereid was discovered by Gerard Kuiper.

1977
Voyager 2 launched on its great journey across the Solar System.

1989
The only space probe to visit, Voyager 2, flew past on 25 August – 12 years after its launch.

1998
The 'Neptune Papers', missing documents from the Royal Greenwich Observatory, were found and further evidence came to light about who actually found Neptune first . The favourite was Le Verrier, but he didn't search for the planet himself. The debate continues to rage on between groups of academics.

2011
Neptune will be in the same position in relation to the Sun as when it was discovered, having completed one orbit.

PLUTO

Since its discovery in 1930, Pluto has held the honour of being the smallest and farthest planet in the Solar System. However discoveries in 2003 and 2005 may threaten this record. A tiny, frozen, distant world, Pluto is the least known of all planets. Our information comes from telescopes only, since no space probe has visited it. Pluto also has a highly unusual orbit, being very oval or 'eccentric'. For part of its immensely long year, Pluto is actually nearer to the Sun than its inner neighbour Neptune.

WHERE IN THE SOLAR SYSTEM?

ATMOSPHERIC CONDITIONS

ATMOSPHERE

Not clearly known, very thin, probably nitrogen, carbon monoxide and methane

NATURE OF SURFACE

Rock and various chemicals frozen as ice

AVERAGE TEMPERATURE

minus 230ºC

LOWEST TEMPERATURE

minus 242ºC

HIGHEST TEMPERATURE

minus 220ºC

WEATHER OR CLIMATE

Atmosphere may move as gases when Pluto is closer to the Sun, but then freeze as nitrogen ice at its farthest distance

SEASONAL CHANGES

As above.

An artist's impression of the surface of Pluto.

PLANET PROFILE

Diameter at Equator **2,274 km**

Surface area **17.9 million sq km**

Tilt of axis **122.5° to its orbit, 115° to orbits of other planets**

Mass (Earth = 1) **0.002**

Volume (Earth = 1) **0.007**

Overall density **1.75 g per cc**

Gravity (Earth = 1) **0.06**

Number of moons **1**

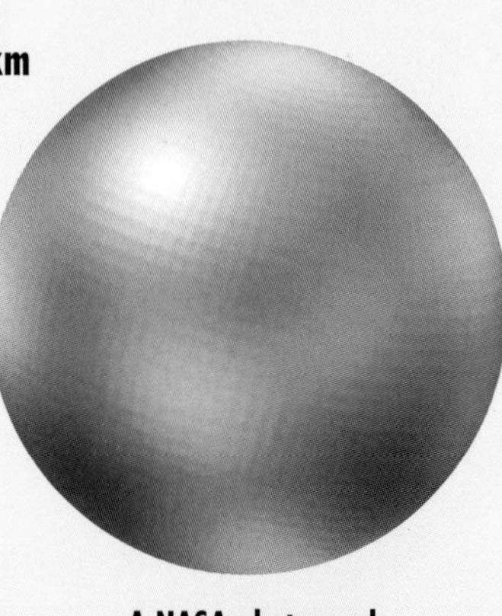

A NASA photograph of Pluto.

ORBIT DETAILS

Average distance from Sun
5,906 million km

Average distance from Sun
39.5 AU (Earth = 1)

Closest distance to Sun (perihelion) 4,436 million km

Farthest distance from Sun (aphelion) 7,375 million km

Average orbital speed
4.67 km per second

Slowest orbital speed
3.68 km per second

Fastest orbital speed
6.11 km per second

Time for one orbit
(Pluto year) 248.1 Earth years

Axial rotation period
(Pluto day) 6.39 Earth days

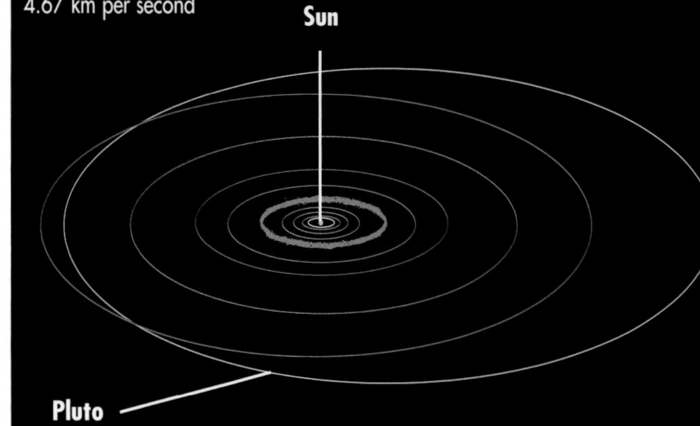

PLUTO'S MOON

Artwork of the planet Pluto and its moon Charon (below) seen from space. Charon, is only half the size of Pluto, which makes the pair a double planet system.

- Charon is the largest moon compared to its planet in the Solar System.
- It measures 1,205 km across, just under half of Pluto's diameter.
- Charon's orbit distance is 19,500 km across and its orbit time is just 6.39 days.
- Charon's spin and orbit time, combined with Pluto's spin, mean that each keeps the same face towards the other at all times. This is called 'tidal locking'.
- Charon's name was officially agreed in 1985 –the 'Ch-' pronounced 'Sh-' as in 'Sharon'.
- This moon has no atmosphere, but its surface is possibly coated in water-ice.

IS PLUTO A TRUE PLANET?

In the 1990s the debate increased about whether Pluto should be called a true planet.

REASONS AGAINST

- Very eccentric (oval) orbit, with closest distance to Sun only 3/5ths of the farthest distance.
- Pluto's orbit is titled at an angle compared to other planets, whose orbits all lie 'flat'.
- Pluto is under half the diameter of the next smallest planet, Mercury.
- Other objects in the Solar System similar in size to Pluto have recently been discovered, most notably an object – provisionally called 'Xena' – which is larger than Pluto. If Pluto is a planet then 'Xena' should be classed as one too.

REASONS FOR

- Pluto is much bigger than any asteroids in the Asteroid Belt and is one of the largest objects to be discovered beyond Neptune.
- It has its own moon and an atmosphere.
- Pluto could be classed as a Trans-Neptunian planet - part of a new class of planets beyond Neptune.

• See pages 34-35 for information on the Asteroid Belt.

An artist's impression of one of the newly discovered Trans-Neptunian planets.

MANY NAMES

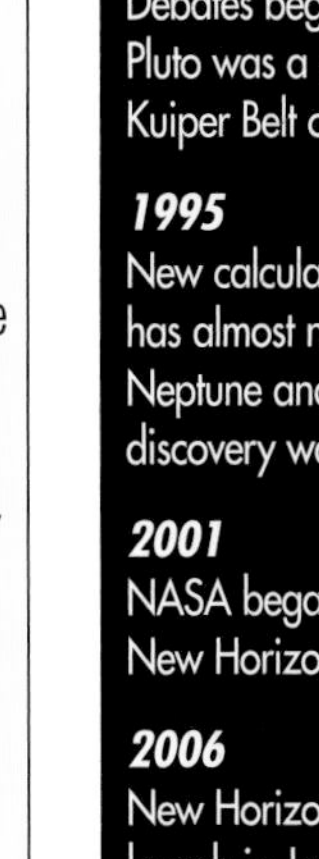

In the week's following Pluto's discovery by Clyde Tombaugh (left), dozens of names were suggested including:

Artemis, Athene, Atlas, Cosmos, Cronus, Hera, Hercules, Icarus, Idana, Minerva, Odin, Pax, Persephone, Perseus, Prometheus, Tantalus, Vulcan and Zymal.

- Pluto was first suggested by Venetia Burney, an 11-year-old girl from Oxford, England.
- She suggested it was so cold and distant, it could be named after the Roman God of the underworld.
- Her grandfather mentioned this to an astronomer friend, who contacted the discovery committee in the USA.
- The name Pluto was quickly agreed, perhaps for one good reason.
- Its first two letters are the initials of Percival Lowell, who founded the observatory where it was discovered.

PLUTINOS

From the 1990s many bodies similar to Pluto, but smaller, have been discovered in the Kuiper Belt beyond Neptune. These Kuiper Belt Objects are called 'Plutinos' if they complete two orbits of the Sun in the same time as Neptune takes to complete three orbits.

SMALLEST PLANET

Pluto is not only the smallest planet, it is smaller in size than 7 moons of some of the giant gas planets Jupiter and Saturn – Ganymede, Titan, Callisto, Io, Europa and Triton. It is even smaller than the Earth's Moon.

ODD ORBIT

- Pluto has by far the most elliptical or 'oval' orbit of any planet.
- From February 1979 to February 1999 Pluto was closer than Neptune to the Sun.

PLUTO TIMELINE

1902
Percival Lowell predicted another body, 'Planet X', beyond Neptune.

1915
Lowell made another prediction, this time fairly close to Pluto's actual size and position, but many experts say that was coincidence (see below).

1930
Pluto was discovered by Clyde Tombaugh at the Lowell Observatory, Arizona.

Pluto became the official name on 30 May.

1978
Pluto's moon Charon was discovered by James Christy.

1977
Voyager 1 was launched and originally was due to visit Pluto, but was redirected to fly past Saturn's moon Titan.

1992
From September hundreds of small, icy objects were discovered beyond Neptune, in a zone now known as the Kuiper Belt.

1993
Debates began as to whether Pluto was a true planet or a Kuiper Belt object.

1995
New calculations showed that Pluto has almost no effect on the orbits of Neptune and Uranus, so its original discovery was largely by chance.

2001
NASA began to plan and build New Horizons space probe.

2006
New Horizons probe planned for launch in January to visit Pluto, Charon and hopefully objects in the Kuiper Belt.

2015
New Horizons expected to fly within 10,000 km of Pluto in July, the first craft to visit the planet, and then within 30,000 km of Charon.

2020
New Horizons may encounter objects in the Kuiper Belt.

ASTEROID TYPES

A stony asteroid

- There are three types of asteroids, made of different materials.
- More than 90% of all known asteroids are called stony asteroids, because they contain stony materials called silicates.
- About another 5-6% of asteroids are made of metal. They contain mostly nickel and iron.
- The rest of the asteroids contain a mixture of silicates and metals.
- Metal asteroids may be from the smashed core of a small planet that was torn apart millions (or billions) of years ago.
- Some asteroids are very dark, because they are rich in carbon compounds.

THE TROJANS

If asteroids stray too close to the giant Jupiter they can get trapped in its orbit. There are two groups of asteroids that circle round and round the Solar System in front and behind Jupiter. Scientists have named these asteroids the Trojans. Sometimes they fall into Jupiter's gravitational pull and become satellites of Jupiter.

An artists' impression of a Trojan asteroid.

ASTEROIDS

Asteroids are chunks of rock that orbit the Sun. They are pieces of rock left over from the formation of the planets and moons. Most asteroids are too far away and too faint to be seen clearly without a telescope. Most orbit far away, beyond Mars, but occasionally one may come closer to the Sun... or Earth. Asteroids have hit the Earth in the past. A major impact about 65 million years ago may be linked to the extinction of the dinosaurs.

ASTEROID FACTS

Asteroids range in size from dust particles to objects nearly 1,000 kilometres across.

- The biggest asteroid, called Ceres, is about 933 kilometres across.
- Ceres was the first asteroid to be discovered. It was found by the Italian astronomer Giuseppe Piazzi in 1801.
- Asteroids spin as they fly through space.
- Large asteroids are tracked in case any of them follow an orbit that may collide with Earth in future.
- The biggest asteroids are ball-shaped, like small planets, leading to their other names – the minor planets or planetoids.

• See page 20 for information on the Moon and its craters.

- Astronomers think the two moons of Mars, Phobos and Deimos, may be captured asteroids.
- The Moon's craters were caused by asteroid impacts.

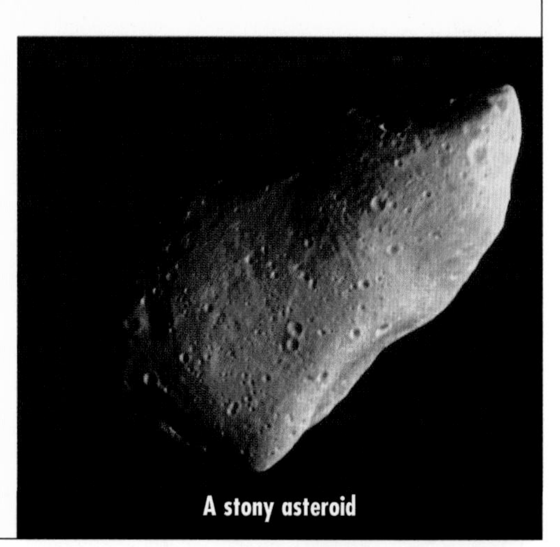

A stony asteroid

WHERE DO ASTEROIDS COME FROM?

- Most asteroids orbit the Sun in a broad band called the Asteroid Belt between the orbits of Mars and Jupiter.
- The Asteroid Belt marks the end of the inner Solar System and the beginning of the outer Solar System.
- Asteroids that cross the orbit of Mars are called Amor asteroids.
- Asteroids that cross the Earth's orbit are called Apollo asteroids.
- Aten asteroids have orbits that are inside the Earth's orbit.
- The Trojans are asteroids that orbit ahead of, or behind, a planet. Mars, Jupiter and Neptune have Trojan asteroids in their orbits.
- There are also rocky and icy bodies orbiting the Sun further out than Neptune in a region called the Kuiper Belt. These are known as Kuiper Belt Object or Trans-Neptune Objects.

• See pages 22-25 for information on Mars and Jupiter

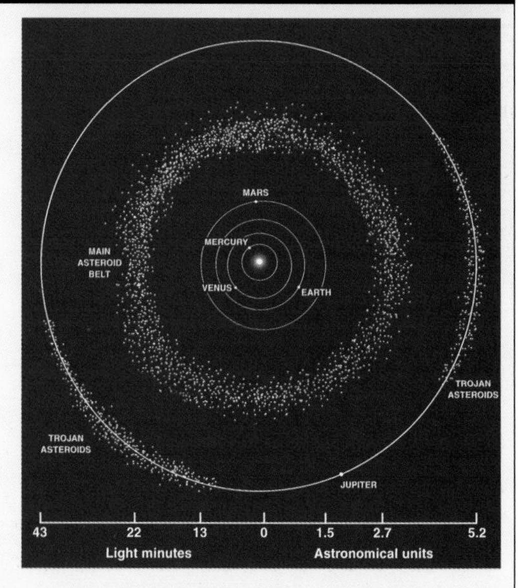

A NASA diagram showing the location of the main asteroid belt.

FIRST DISCOVERIES

Below is a list of the first ten asteroids discovered.

Asteroid	Discovered	Size (km)
Ceres	1801	960x932
Pallas	1802	571x525x482
Juno	1804	288x230
Vesta	1807	525
Astraea	1845	120
Hebe	1847	204
Iris	1847	208
Flora	1847	162
Metis	1848	158
Hygeia	1849	430

Ceres was discovered by the Italian astronomer Giuseppe Piazzi.

STRANGEST ASTEROIDS

- Icarus is the asteroid with the strangest orbit. It comes closer to the Sun than the planet Mercury.
- The smallest asteroids are all sorts of odd shapes, because their pull of gravity is not strong enough to pull them into a ball shape.
- The asteroid with the strangest shape seen so far is probably Kleopatra. It is a 220 kilometres long chunk of rock in the shape of a dog's bone!

ASTEROIDS WITH MOONS

- Asteroids have a very weak pull of gravity, but they are still able to attract and capture smaller asteroids as moons.
- The asteroid Ida has a small moon called Dactyl. Ida is 56 km across and its moon, Dactyl, is only about 1km across.
- An asteroid called 45 Eugenia may have a small moon too.
- Some asteroids travel in pairs, called binaries, that orbit each other.
- An asteroid called 4179 Toutatis is thought to be two asteroids, one 2.5km across and the other 1.5km across, that may be actually touching each other.

Nine photographs of Ida taken by the probe Galileo.

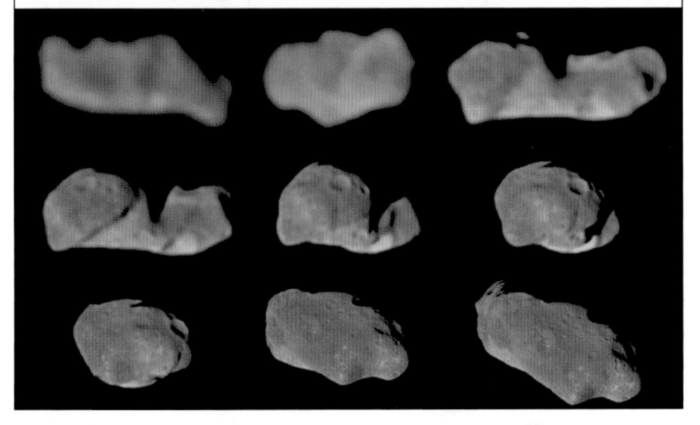

NEAR DISASTERS

In 1908 an asteroid slammed into a place called Tunbuska in Siberia, flattening trees in an area 100 km across. If this happened today the effect would be devastation. In 1991 there was a close shave when a small asteroid passed just 170,000 km over the Earth. This is close in astronomical terms! Today, astronomers are looking for near-Earth objects like these. If they find them, there are various theories as to what they might be able to do to stop the threat. Some people think a nuclear missile could be launched into space to nudge the asteroid to one side.

SPACE PROBES TO ASTEROIDS

A series of probes have been sent into space to learn more about asteroids. Here is a list.

Space probe	Details
Galileo	On its way to Jupiter, Galileo passed asteroid Gaspra in 1991 and Ida in 1993.
Hayabusa	Met with asteroid 25143 Itokawa and due to return particles from it to Earth in 2007.
NEAR-Shoemaker	Passed asteroid Mathilde on its way to 433 Eros. Orbited Eros 230 times and then landed – the first spacecraft ever to make a controlled landing on an asteroid.
Deep Space 1	Flew past asteroid 9969 Braille (formerly known as 1992KD) on its way to a comet.

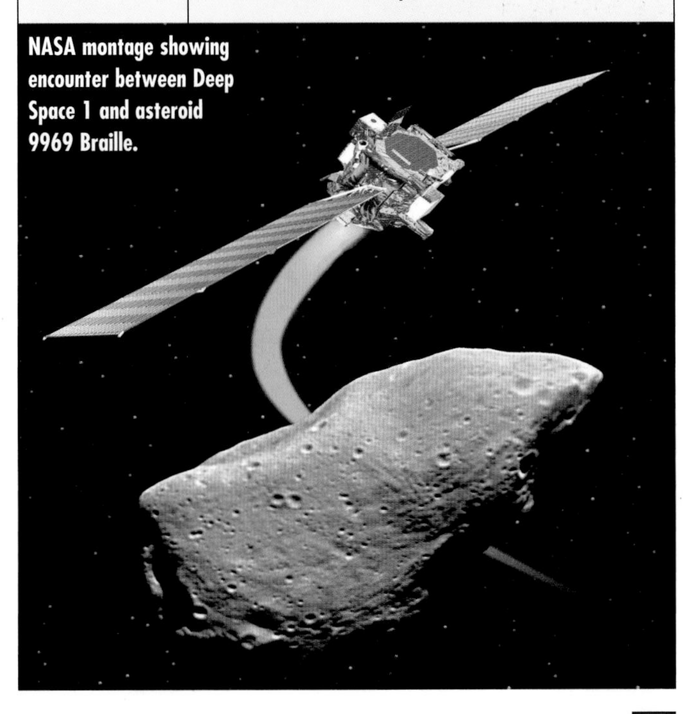

NASA montage showing encounter between Deep Space 1 and asteroid 9969 Braille.

METEORS

Look up into a clear sky on any night and you might be lucky enough to see a streak of light. It appears for only a fraction of a second and then it is gone. The bright streak is called a meteor, and is made by a particle of dust entering the Earth's atmosphere from space and burning up. Large meteors that survive their journey down through the atmosphere and hit the ground are called meteorites.

METEORITE FACTS

- About 500 baseball-sized rocks from space hit the ground every year.
- The largest meteorites hit the ground with such force that they make a hole called an impact crater.
- The biggest meteorite ever found is called the Hoba meteorite (after the place in Namibia where it landed). It weighs about 60 tonnes!
- A meteorite from Mars, called ALH84001, caused a stir in 1996, when some scientists thought they had found evidence of life inside it. However, most scientists now think the features seen in the meteorite are not signs of life.

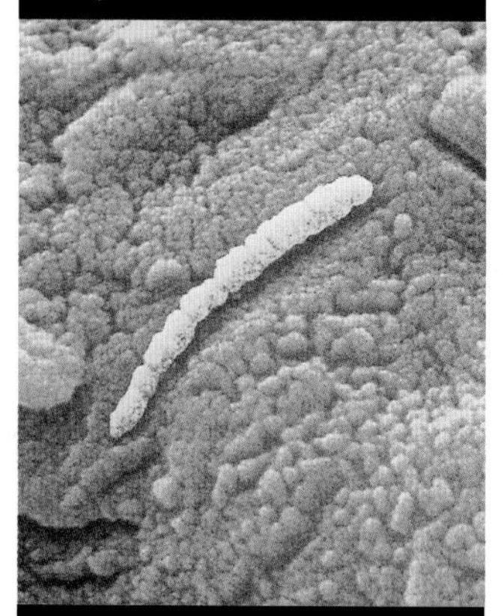

Close-up of structure found in meteorite from Mars.

CRATERS

Large meteorites have struck Earth and left giant craters. One of the most famous of these is the Barringer Crater in Arizona (shown below), which measures over 1.2 km across.
It was created 50,000 years ago by the impact of a 50-metre-wide meteorite.

METEORS

A spectacular meteor shoots to Earth.

Meteors are also called shooting stars, although they are not stars.

- Meteors range in size from a grain of sand to a tennis ball.
- Some meteors called bolides explode in the atmosphere with a sound like thunder.
- Meteoroids, the particles that become meteors, enter the Earth's atmosphere at up to about 70 kilometres per second (250,000 kilometres per hour.
- Scientists estimate that up to 4 billion meteors streak through the Earth's atmosphere every day.

METEOR SHOWERS

There are a lot more meteors than usual at certain times of the year. These events are called meteor showers.

- Meteor showers occur when the Earth flies through a trail of particles left behind by a comet.
- The Earth passes through a comet's tail in the same part of every orbit, so meteor showers occur at the same times every year.
- All the meteors in a meteor shower appear to come from the same point in the sky.
- Meteor showers are named after the constellation in the direction the meteors appear to come from.
- Some meteor showers can produce hundreds of shooting stars an hour. Meteor showers last for anything from a few hours to several days.

• *See pages 18-19 for information on Earth.*

A NASA photograph showing a spectacular meteor shower.

TYPES OF METEORITES

The three main types of meteorites are called irons, stony meteorites and stony-irons.

- Irons are made of an alloy (mixture) of iron and nickel.
- Stony-irons contain rock and iron-nickel alloy.
- There are three types of stony meteorites, called chondrites, carbonaceous chondrites and achondrites.
- Chondrites are made of small ball-shaped particles called chondrules, made of minerals that have melted and fused together.
- Chondrites may be the oldest rocks in the Solar System.
- Carbonaceous chondrites contain carbon.
- Achondrites are meteorites made from stone but without the spherical chondrules found in chondrites.
- Most meteorites are chondrites.
- Achondrites may be rocks blasted out of the surface of the Moon or Mars by asteroid impacts.

stony irons

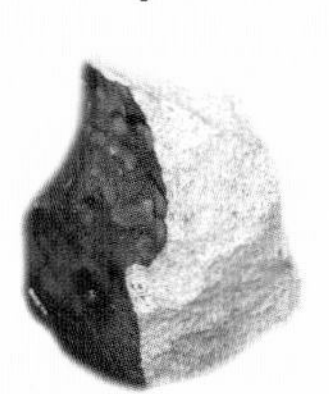

stony meteorite

iron meteorite

BEST METEOR SHOWERS

- Perseids (below) is named after the constellation Perseus. It can be seen between July 23 and August 22.
- Orionids is named after the constellation Orion. It can be seen between October 15-29.
- The meteor shower Geminids is named after the Gemini constellation. It can be seen in the sky between December 6-19.

THE TEN BIGGEST METEORITES

The largest meteorite ever found is to be found in the grounds of the ``Hoba" farm close to Grootfontein, Namibia.

Where found	When found	Weight
Hoba, Namibia	1920	60 tonnes
Campo del Cielo, Argentina	1969	37 tonnes
Cape York, Greenland	1894	31 tonnes
Armanty, Outer Mongolia	1898	28 tonnes
Bacuberito, Mexico	1863	27 tonnes
Mbosi, Tanzania	1930	26 tonnes
Cape York, Greenland	1963	20 tonnes
Willamette, Oregon, USA	1902	14 tonnes
Chupaderos, Mexico	1852	14 tonnes
Mundrabilla, Australia	1966	11 tonnes

PARENT COMETS

A comet whose trail produces a meteor shower is called the shower's parent comet.

Meteor shower	Parent comet
Eta Aquarids	Halley's Comet
Geminids	Asteroid 3200 (Phaethon)
Leonids	Tempel-Tuttle
Lyrids	Thatcher
Orionids	Halley
Perseids	Swift-Tuttle
Taurids	Encke
Ursids	Tuttle

The spectacular Eta Aquarids shower photographed in 1987.

COMETS

Every few years, an object that looks like a fuzzy star with a long bright tail appears in the sky. These strange objects are not stars. They are comets. A comet is a chunk of gas and dust and ice left over from the formation of the solar system. Comets orbit the Sun. When a comet nears the Sun, some of the ice on its surface evaporates and releases dust to form the tail. Most comets are too dim to be seen with the naked eye, but every ten years or so an especially bright comet appears in the sky.

COMET FACTS

- About 850 comets have been spotted and listed by astronomers.
- Comets are named after their discoverers.
- Cometary nuclei (the centre of comets) are typically only a few tens of kilometers across, although many are much smaller.
- If all the known comets were added together, they would weigh less than the Moon.
- A comet's tail always points away from the Sun.
- The idea that a comet is made of dust and ice, like a dirty snowball, was suggested Fred Whipple in 1950.
- If the Earth passes through the trail of dust particles left behind by a comet, you may see lots of meteors (shooting stars) as the dust enters the Earth's atmosphere and burns up.

The astronomer Fred Whipple pictured in 1986.

FAMOUS COMETS

- Halley's Comet reappears every 76 years. It is named after the astronomer Edmond Halley.
- Hale-Bopp is a long-period comet that is seen only once every few thousand years.
- Encke has the shortest period of all comets – it reappears every 3.3 years.

• See page 11 for information on Edmond Halley.

THE STRUCTURE OF A COMET

A comet has three parts – the nucleus, the coma and the tail.

- The nucleus is the solid part of a comet, in the middle of the comet's fuzzy head.
- The coma is the gassy atmosphere that surrounds the nucleus when the comet nears the Sun.
- The tail is the long bright trail of dust and gas that streams away from the comet.
- A comet's dust tail can be as much as 10 million kilometres long.
- The coma and tail look bright only because they reflect sunlight.

WHERE DO COMETS COME FROM

Comets originate from two places in the Solar System.

- A belt of icy objects, called the Kuiper Belt, begins at about the orbit of Neptune and stretches beyond the orbit of Pluto.
- A ball-shaped cloud of icy objects, called the Oort Cloud, surrounds the whole solar system.
- The Oort Cloud lies about 12 trillion kilometres from the Sun.
- Scientists think the Oort Cloud may contain 10 trillion comets (that's 10 followed by 12 noughts).
- The Oort Cloud is named after the Dutch astronomer, Jan Hendrik Oort (1990-1992), who suggested the idea of the distant cloud of comets in 1950.

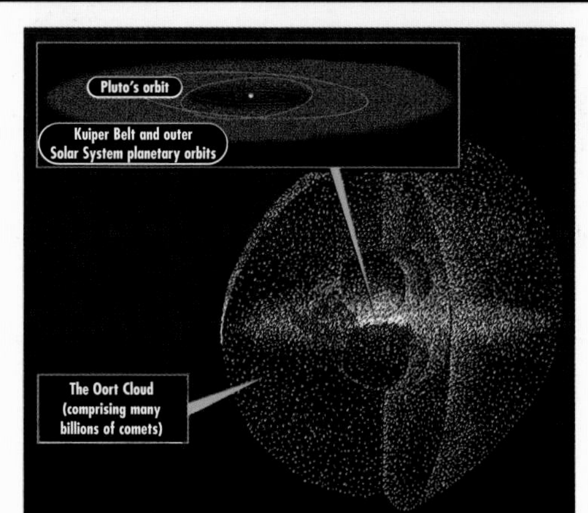

A diagram showing (inset) the location of the Kuiper Belt in the Solar System, and (main picture) the Oort Cloud, made up of billions of comets.

• See pages 30-33 for information on Neptune and Pluto.

COMET ORBITS

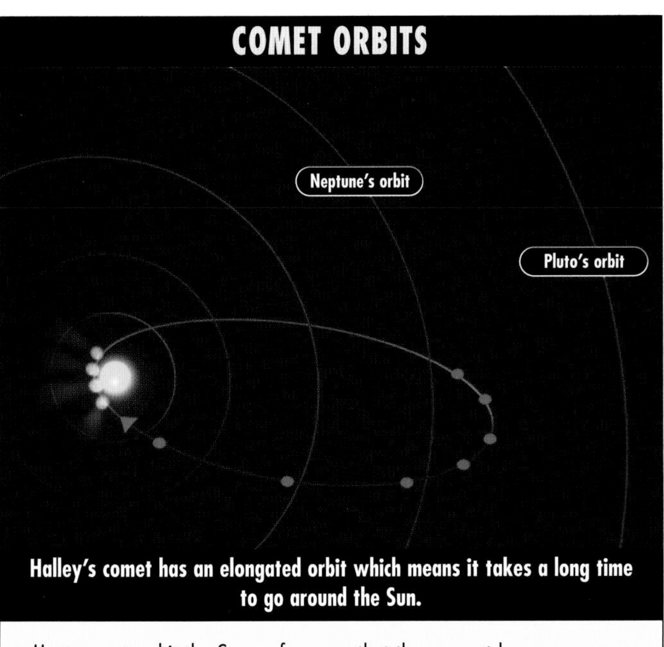

Halley's comet has an elongated orbit which means it takes a long time to go around the Sun.

- Most comets orbit the Sun so far away that they cannot be seen.
- A passing planet can deflect a comet out of the Kuiper Belt and send it on a new orbit closer to the Sun.
- Oort Cloud comets are so far away that passing stars can tug them into a new orbit that takes them towards the inner solar system.
- Comets from the Kuiper Belt orbit the Sun faster than other comets. They are called short-period comets.
- Comets from further away in the Oort Cloud take longer to orbit the Sun and so they are called long-period comets.

COMET HISTORY

People in the ancient world feared comets as signs of coming disasters and recorded their sightings. Even recently, sightings are greeted with a sense of wonder.

- The 'hairy star' that appears on the Bayeux Tapestry is thought to be Halley's Comet. The tapestry shows the Norman conquest of England in 1066.
- As recently as 1910, some people in Chicago were reported to have boarded up their windows to protect themselves from Halley's Comet!
- Records of Halley's Comet date back to 240BCE with certainty, and perhaps as far back as 1059BCE.
- The Great Comet of 1843 was probably the brightest comet ever seen. It was clearly visible in daylight.

Halley's comet (top left) depicted on the Bayeux Tapestry.

SOME SPACE PROBES TO COMETS

Space probe	Launched	Comet
ICE (International)	12/8/78	Flew past comet Giacobini-Zinner
Vega 1 (Russia)	15/12/84	Flew past comet Halley
Vega 2 (Russia)	21/12/84	Flew past comet Halley
Sakigake (Japan)	8/1/85	Flew past comet Halley
Giotto (Europe)	2/7/85	Flew close to comet Halley and photographed its nucleus
Suisei (Japan)	18/8/85	Flew past comet Halley
Stardust (USA)	7/2/99	Flew past comet Wild 2 and collected particles for return to Earth in 2006
Rosetta (Europe)	2/3/04	Due to rendezvous with comet Churyumov-Gerasimenko in 2014
Deep Impact (USA)	21/12/04	Flew past comet Tempel 1 and smashed a mini-probe into it

The Deep Impact spacecraft lifts off from Launch Pad 17-B, Cape Canaveral Air Force Station.

STARS

A star is a giant ball of glowing gas in space fuelled by nuclear reactions in its core. You can see several thousand stars with the naked eye. But these are only the brightest stars. Astronomers have found tens of millions more stars by using powerful telescopes to probe the sky. Our star, the Sun, is an ordinary star. Compared to the Sun, some stars are giants. They each contain enough matter to make tens or hundreds of Suns.

STAR BRIGHTNESS

- How bright a star appears to be depends on how bright it really is and how far away it is.
- The closer a star is to Earth, the brighter it looks.
- Astronomers call a star's brightness its magnitude.
- A star's magnitude (brightness) is given by a number.
- The brighter a star is, the smaller or lower its magnitude is.
- Stars of magnitudes +6 or more are too faint to be seen with the naked eye.
- The Sun is the brightest object in the sky, with a magnitude of –26.8.

• See pages 12-13 for information on the Sun.

STARS TOGETHER

- Most stars are not single stars like the Sun. They have at least one companion star.
- The two stars orbit each other.
- Some pairs of stars look close together only because they lie in the same direction from Earth, but their movements show that they are not orbiting each other.
- Sometimes two stars are so close together that one star sucks gas off the other star.
- Extra gas falling on a star may explode in a giant blast called a nova.

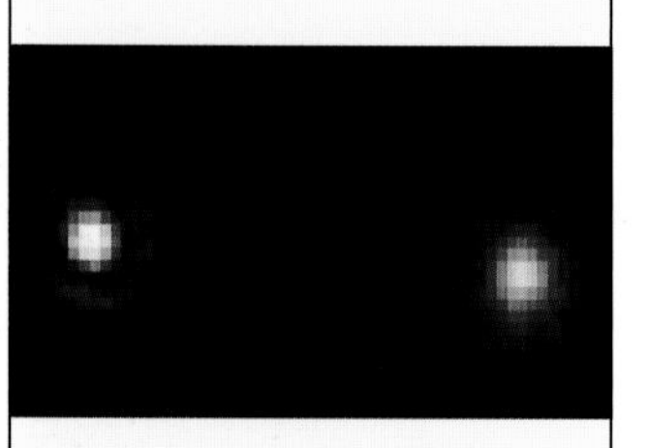

Mira A (right) with its companion star on the left.

NAMES OF STARS

Artwork from an Egyptian temple showing the signs of the Zodiac at the centre.

• See pages 8-9 for information on early astronomers.

- Some of the names we know stars by today were given to them by ancient Greek and Arab astronomers perhaps two thousand years ago.
- In the 17th century, the German astronomer Johann Bayer started naming stars using Greek letters.
- The brightest star in a constellation was called alpha, the next brightest beta and so on. For example, the brightest star in the constellation Centaurus is called Alpha Centauri.
- When Greek letters ran out, astronomers named lots of fainter stars by adding 'ordinary' numbers to their constellation name. For example, a faint star in Pegasus is called 51 Pegasi.
- Today, new stars are identified by numbers, often with the name of the person who discovered them and the year of discover.
- The Hubble Space Telescope is pointed in the right direction using a catalogue of 15 million stars whose positions are known with great accuracy by astronomers.

BRIGHTEST STARS

After the Sun, the ten brightest stars seen from Earth are:

STAR	CONSTELLATION	MAGNITUDE
Sirius	Canis Major	-1.44
Canopus	Carina	-0.62
Arcturus	Bootes	-0.05
Alpha Centauri A	Centaurus	-0.01
Vega	Lyra	0.03
Capella	Auriga	0.08
Rigel	Orion	0.18
Procyon	Canis Minor	0.40
Achernar	Eridanus	0.45
Betelgeuse	Orion	0.45

The star Sirius (brightest object) in the constellation Canis Major.

COLOURS AND HOTNESS

Stars are different colours. The colour of a star shows its temperature (how hot it is).

STAR TYPE	STAR COLOUR	TEMPERATURE
O	Blue	over 25,000C
B	Blue	11,000-25,000C
A	Blue	7,500-11,000C
F	Blue-white	6,000-7,500C
G	White-yellow	5,000-6,000C
K	Orange-red	3,500-5,000C
M	Red	Under 3,500C

CLOSEST STARS

After the Sun, the ten closest stars to Earth are:

STAR	CONSTELLATION	DISTANCE
Proxima Centauri	Centaurus	4.2 light years
Alpha Centauri	Centaurus	4.4 light years
Barnard's Star	Ophiucus	6.0 light years
Wolf 359	Leo	7.8 light years
Lalande 21185	Ursa Major	8.3 light years
Sirius	Canis Major	8.6 light years
Luyten 726-8	Cetus	8.7 light years
Ross 154	Sagittarius	9.7 light years
Ross 248	Andromeda	10.3 light years
Epsilon Eridani	Eridanus	10.5 light years

BIRTH AND DEATH OF A STAR

All stars are born from clouds of dust, and end their lives in violent circumstances. They begin life as dwarfs before changing into giants or supergiants as they heat up. Depending on how much mass they start out with, they end their lives in a variety of different ways.

THE BIRTH OF A DWARF STAR

1. New stars come from giant clouds of dust and gas.

2. Knots begin to form in the gas cloud as gravity pulls it together. This compression causes the cloud to heat up.

3. Eventually the gas begins to spiral round. Jets of gas are expelled from the poles.

4. The star's brightness increases as nuclear fusion begins at its centre. All the gas and dust in the space surrounding the star is blown away, and eventually the star emerges from its dusty cocoon.

5. The process is complete. The new dwarf star begins to shine.

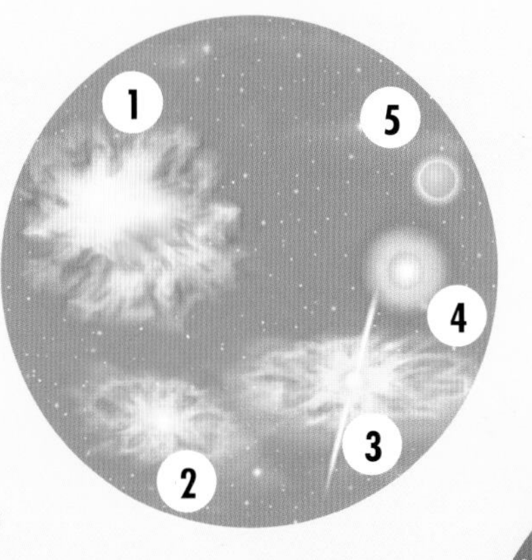

THE DEATH OF A SUPERGIANT

1. Like a dwarf, a supergiant starts its life in the same way, but it is much hotter and brighter and can be hundreds of times the diameter of the Sun.

2. After a brilliant but short career, a supergiant commits cosmic suicide in a spectacular explosion called a supernova.

3. A supernova leaves behind an extremely dense remnant such as a neutron star or a black hole.

1
2
3
neutron star
black hole

THE DEATH OF A GIANT

1. If a star uses all the hydrogen in its central core, hydrogen burning will start to occur in the surrounding shells, which then become heated and cause the outer envelope of the star to swell outwards.

2. As a giant's interior gets hotter and hotter, eventually it puffs away its bloated outer shell. This is called a planetary nebula.

3. The hot remnant left behind after a giant has passed the planetary nebula stage is called a white dwarf. The gravity of white dwarfs is so intense that the result is an Earth-sized remnant so dense that a matchboxful would weigh several tonnes.

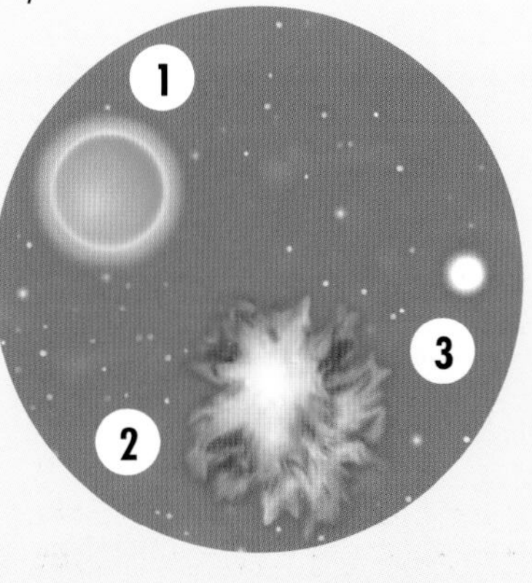

Giants

A giant is a former dwarf that has cooled and expanded to a great size. In 5,000 million years from now, this will be the fate of our Sun.

Supergiants

A supergiant starts its life as a main sequence dwarf, but it is much brighter, hotter and massive than the Sun, and can be hundreds of times bigger. It has a hectic but very short life.

HOW MANY CONSTELLATIONS?

- There are a total of 88 constellations.
- The ancient Egyptian astronomer, Ptolemy, listed 48 constellations in his book, *The Almagest*, written in AD150.
- European astronomers added another 40 constellations in the 17th and 18th centuries.
- During the early 1900s, the International Astronomical Union mapped the boundaries between all the constellations so that there were no gaps between the constellations and so every star now belongs to a constellation.

• See pages 8-11 for information on astronomers.

FINDING NAMES

- Most constellations were named according to religious beliefs and mythological characters.
- The oldest constellations were probably named more than 4,000 years old.
- The ancient Greeks had no names for constellations in the southernmost skies, because the stars there were not visible from Greece.
- European astronomers filled in the gaps between the northern constellations and began naming the southern constellations.
- The final gaps in the southern constellations were filled by the French astronomer Nicolas Louis de Lacaille.

STAR CONSTELLATIONS

People have seen patterns in the stars since ancient times. The groups of stars that form these patterns are called constellations. The stars in a constellation rarely have any connection with each other. They simply lie in the same direction when viewed from Earth. Twelve of the most ancient constellations have special significance. They are the twelve constellations the Sun, Moon and brightest planets pass through. They are also known as the signs of the Zodiac. The ancient constellations are still used by astronomers as signposts to find their way around the night sky.

Winter sky

Orion, the Hunter, is a magnificent constellation visible during late evenings in winter. The three stars in its belt can be used as a celestial signpost. Just below the belt is a shiny patch called the Orion Nebula, which is a splendid sight through binoculars or a small telescope. The Orion Nebula is in fact a stellar nursery, where stars are being born right now.

GEMINI

A slightly curving line drawn upwards through Rigel and Betelgeuse will get you to Gemini, with its two bright stars Castor and Pollus, the Heavenly Twins.

AURIGA

Over Orion's head is Auriga, the Charioteer. Near the bright star Capella is a distinctive triangle of stars called the Kids.

PERSEUS

Now follow a line northeast of Orion past Taurus and you will come to Perseus. This constellation contains a double open cluster, which is a great sight through binoculars.

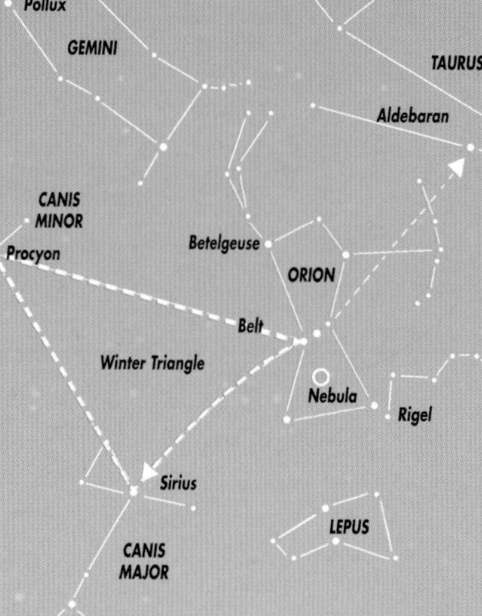

CANIS MINOR

A line to the west of Orion takes you to the small constellation Canis Minor, the Little Dog. The three stars Procyon (in Canis Minor), Betelgeuse (in Orion) and Sirius (in Canis Major) form the prominent Winter Triangle.

TAURUS

Follow the three stars of Orion's belt upwards and you will come to the constellation Taurus, the Bull. Taurus contains the bright red star Aldebaran. This star appears to form part of the 'v' of the Hyades, which is an open star cluster. In fact, Aldebaran is a foreground star and is not part of this distant group. Following the line from Orion's belt yet further, you will come to a close-knit bunch of stars called the Pleiades. These stars form yet another open cluster.

CANIS MAJOR

Canis Major, the Big Dog, is found by following Orion's belt downwards. It contains Sirius, the brightest star in the sky.

LEPUS

Beneath Orion is an undistinguished constellation called Lepus, the Hare.

ERIDANUS

Eridanus, the River, is another faint constellation which manages to meander a sixth of the way around the sky. It lies to the right of Orion, just past Rigel.

ORION

- Orion is one of the oldest constellations.
- More than a thousand years ago it was known as Tammuz to the Chaldeans.
- Tammuz was the name of the month when the three stars across its middle rose before sunrise.
- The Syrians knew it as Al Jabbar (The Giant).
- The ancient Egyptians knew it as Sahu, the soul of the god Osiris.
- The name we know it by (Orion) comes from the ancient Greeks.
- In Greek mythology Orion was a giant hunter.

The constellation Orion.

Spring sky

When you look up at the late evening sky in spring, you should be able to see the seven stars of the Plough. Use the Plough (marked below in red) to navigate your way around the sky.

CEPHEUS

A straight line through Merak and Dubhe in the Plough and Polaris will take you to Cepheus, a dim constellation.

CASSIOPEIA

A line from Mizar in the Plough through Polaris takes you to the constellation Cassiopeia, a beautiful w-shaped constellation through which parts of the Milky Way pass.

DRACO

Between Ursa Major and Ursa Minor is long, winding Draco, the Dragon, a fairly dim constellation.

URSA MINOR

Follow the two stars Merak and Dubhe in the Plough northwards and you will come to Polaris, the Pole Star, in the constellation Ursa Minor, the Little Bear.

BOOTES

The three left-hand stars of the Plough can be used to trace a gentle curve downwards to the bright orange star Arcturus in the constellation Bootes, the Herdsman.

URSA MAJOR & THE PLOUGH

The Plough, or Seven Stars, is not actually a constellation but the brightest part of the constellation Ursa Major, the Big Bear. The most important thing about the Plough is that some of its stars make useful signposts to other parts of the sky.

LEO

Directly underneath the Plough is the constellation Leo, the Lion. It is one of the few constellations that bears even the slightest resemblance to its name. Its bright star, Regulus, is the dot in an inverted question mark of stars known as the Sickle.

LARGEST CONSTELLATIONS

1. Hydra (The Sea Serpent)
2. Virgo (The Virgin)
3. Ursa Major (The Great Bear)
4. Cetus (The Sea Monster)
5. Hercules (Hercules)

SMALLEST CONSTELLATIONS

1. Crux (The Southern Cross)
2. Equuleus (The Little Horse)
3. Sagittarius (The Archer)
4. Circinus (The Drawing Compass)

SIGNS OF THE ZODIAC

The twelve constellations of the Zodiac and the dates when the Sun passes through them are:

CONSTELLATION	DATES
Aries (The Ram)	21/03 – 19/04
Taurus (The Bull)	20/04 – 20/05
Gemini (The Twins)	21/05 – 21/06
Cancer (The Crab)	22/06 – 22/07
Leo (The Lion)	23/07 – 22/08
Virgo (The Virgin)	23/08 – 22/09
Libra (The Balance)	23/09 – 23/10
Scorpius (The Scorpion)	24/10 – 21/11
Sagittarius (The Archer)	22/11 – 21/12
Capricornus (The Goat)	22/12 – 19/01
Aquarius (The Water Bearer)	20/01 – 18/02
Pisces (The Fish)	19/02 – 20/03

BRIGHTEST CONSTELLATIONS

The ten brightest stars lie in the following constellations:

Star	Constellation
Sirius	Canis Major
Canopus	Carina
Arcturus	Boötes
Alpha Centauri	Centaurus
Vega	Lyra
Capella	Auriga
Rigel	Orion
Procyon	Canis Major
Achernar	Eridanus
Betelgeuse	Orion

TELESCOPES

Astronomers have used telescopes to study the sky for about 400 years. The first telescopes magnified images of distant objects by means of lenses. Later, a new type of telescope had curved mirrors instead of lenses. Astronomers use telescopes to collect all sorts of waves and rays, including visible light, from space to learn more about stars, galaxies and other objects.

THE WORLD'S BIGGEST TELESCOPE

- The Hobby-Eberly Telescope in Texas, USA, has the largest primary mirror ever made. It measures 11.1 metres by 9.8 metres.
- The biggest steerable radio telescope is a 100-metre dish at Effelsberg, in Germany.
- The biggest radio telescope is at Arecibo in Puerto Rico. The 305-metre dish was built in 1963.
- Ten 25-metre radio telescopes called the Very Long Baseline Array, located from Canada to Hawaii, work together to produce very detailed radio pictures of the sky.

TYPES OF LIGHT

Light is a form of radiation that is transported in waves. Each of the colours of the rainbow has its own wavelength. The entire range of wavelengths is called the electromagnetic spectrum. Unfortunately, the protective atmosphere of the Earth cuts out many of the wavelengths, but from space, the entire spectrum is visible. By studying what kind of radiation is emitted from objects such as stars, astronomers can learn about an object's density, temperature, chemical composition and how it moves.

visible light
radio waves
infra-red
X-ray
microwaves
ultra violet
gamma rays
wavelength increases

TELESCOPE PARTS

- A refracting telescope, or refractor, collects light with a large objective lens. The image is viewed through a smaller eyepiece lens.
- A reflecting telescope collects light with a large primary, or main, mirror and reflects it to an eyepiece, cameras or other instruments by means of a smaller secondary mirror.
- Large mirrors sag under their own weight, so the biggest reflecting telescopes have mirrors made in smaller sections joined together.
- Each of the two Keck telescopes on Mauna Kea, Hawaii, has a 10-metre primary mirror made from 36 segments.

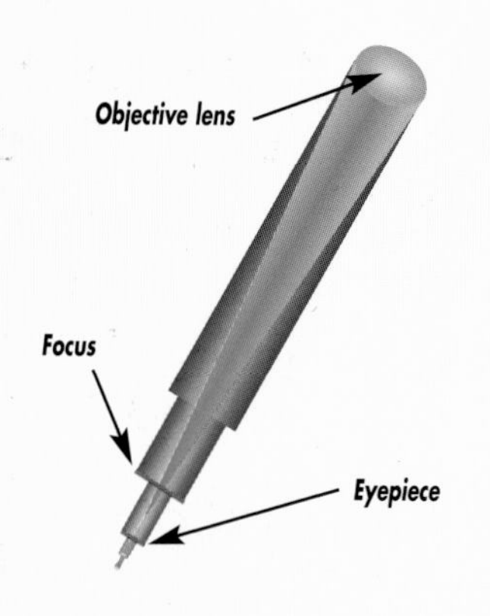

REFRACTING TELESCOPE

1. The objective lens catches the light and brings it to a focus.

2. The eyepiece magnifies the focused image.

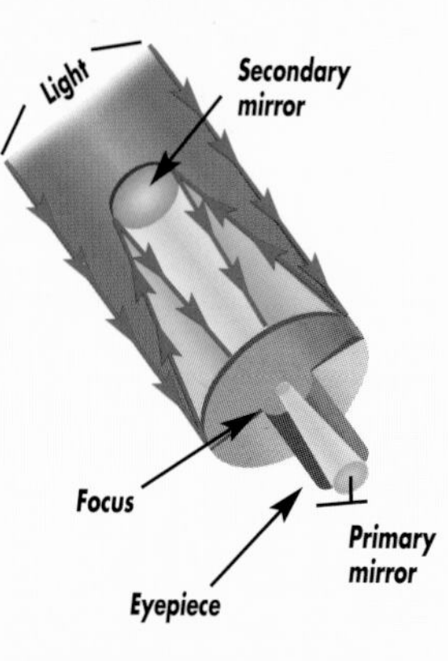

REFLECTING TELESCOPE

1. Light falls through the top of the open-frame tube, and heads towards the primary mirror.

2. It is then reflected up the tube to the smaller, secondary mirror.

3. The light is then reflected back down the tube, through a hole in the primary, to the focus.

TYPES OF TELESCOPE

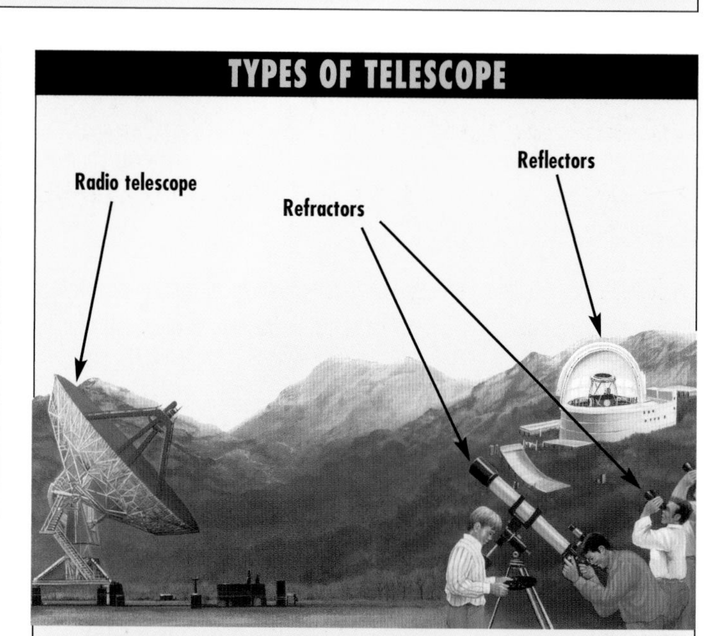

Telescopes can be divided or classified in different ways. They can be divided according to the way they produce a magnified image.

- Refractors use lenses to form an image.
- Reflectors use curved mirrors to form an image.
- Refractors and reflectors collect and magnify light, but telescopes can make use of other types of energy.
- Radio telescopes use radio waves; infra-red telescopes use infra-red rays; X-ray telescopes use X-rays; ultra-violet telescopes use UV waves and gamma ray telescopes use gamma rays.

• See pages 10-11 for information on later astronomers.

MAKING SHARP IMAGES

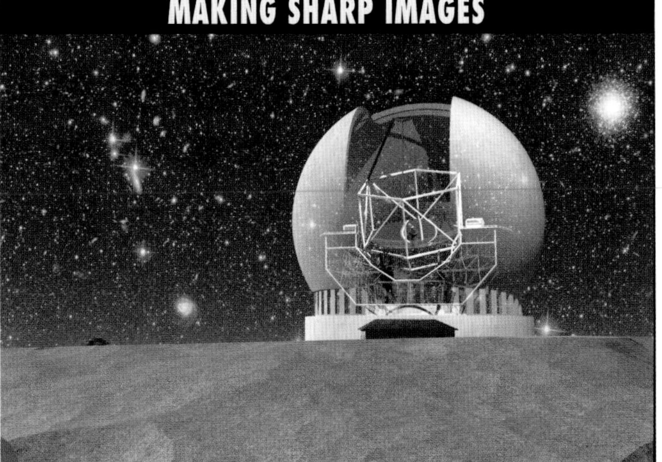

The California Extremely Large Telescope in the United States.

- Starlight is distorted as it travels down through the Earth's atmosphere to the ground.
- The distorting effect of the atmosphere makes stars appear to twinkle, and also make it difficult for astronomers to take clear photographs of them.
- Astronomers avoid the distorting effect of the atmosphere by placing their telescopes on top of mountains, above the thickest part of the atmosphere.
- Telescopes in space, above all of the Earth's atmosphere, have the clearest view of all.
- Some reflecting telescopes have 'adaptive optics' – the mirror continually flexes (changes shape) to cancel the effect of the atmosphere and produce sharper images.
- Some telescopes work only in space, because the energy they use to make images is blocked by the Earth's atmosphere.

SPACE TELESCOPES

TELESCOPE	LAUNCHED	USED TO STUDY
Beppo-Sax	1996	X-rays
Chandra X-ray Observatory	1999	X-rays
Compton Gamma Ray Observatory	1991	gamma rays
Einstein (HEAO-2)	1978	X-rays
Hipparcos	1989	star positions
Hubble Space Telescope	1990	stars, galaxies, nebulae
Infra-red Astronomical telescope	1983	infra-red waves
International Space Observatory	1995	infra-red waves
International Ultraviolet Explorer	1978	ultraviolet waves
Solar and Heliospheric Observatory	1995	the Sun
Spitzer Space Telescope	2003	infra-red waves
Uhuru	1970	X-rays

LINKING TELESCOPES

The amount of detail a telescope can see depends on how much light or other energy it collects.

- When several telescopes are pointed in the same direction and linked together, they behave like a bigger telescope.
- The twin Keck telescopes in Hawaii have mirrors 10 metres across, but when they are linked together, they behave like one telescope with a mirror 85 metres across.
- The European Very Large Telescope (VLT) has four mirrors, each 8.2 metres across. When linked together, they mimic a telescope with a 200-metre mirror.

TELESCOPES OF THE FUTURE

- The Giant Magellan Telescope will be nearly 22 metres across and it will have seven huge 8.2-metre mirrors.
- The Thirty Metre Telescope, with a mirror 30 metres across, will be sited in Hawaii, Chile or Mexico.
- The European large Telescope (ELT) will have a mirror 100 metres across, about ten times the size of the biggest telescope mirrors today.
- In 2007, the Herschel Space Observatory will be launched to study infra-red and sub-millimetre waves.
- The Hubble Space Telescope's successor, the James Webb Space Telescope will be launched in about 2013.

TELESCOPE TIMELINE

1610
Galileo Galilei becomes the first person to study the night sky through a telescope.

1668
Sir Isaac Newton builds the first reflecting telescope.

1845
Lord Rosse builds a giant 72-inch telescope at Birr Castle in Parsonstown, Irleand.

1895
The Yerkes 40-inch refractor is built in Williams Bay, Wisconsin, USA.

1908
The Mount Wilson 100-inch reflecting telescope first used in California.

1937
Karl Jansky builds the first radio telescope.

Grote Reber builds a 31-foot radio telescope.

1957
The 250-foot Jodrell Bank steerable radio telescope is completed.

1963
The 305-metre Arecibo radio telescope in Puerto Rico begins operating.

1974
The 153-inch Anglo-Australian telescope in Australia opens.

1979
The 150-inch UKIRT, 140-inch optical reflector and NASA Infra-Red Facility begin work on Mauna Kea, Hawaii.

1990
The Hubble Space Telescope is launched by the Space Shuttle.

1993
The first 10-metre Keck telescope begins work at Mauna Kea, Hawaii.

1996
Keck II begins operating on Mauna Kea, Hawaii.

2003
The Spitzer Space Telescope is launched in August.

MILKY WAY

Our star, the Sun, is one of billions of stars that travel through space together. This vast collection of stars is called the Milky Way galaxy. On a clear dark night, you may be able to see the hazy band of the Milky Way stretching across the sky. The stars are held together by the pull of their gravity.

SIZE AND SHAPE

- The Milky Way is a thin disc of stars with a thicker bulge in the middle.
- It is sometimes described as looking like two fried eggs back to back!
- Stars are packed more closely together in the central bulge than in the rest of the disc.
- Stars are not spread evenly across the disc. They form arms that curl away from the centre in a spiral shape.
- A beam of light would take about 100,000 years to cross the Milky Way from one side to the other. So we say the Milky Way measures 100,000 light-years across.
- The disc is surrounded by a ball-shaped 'halo' of globular clusters containing very old stars.
- The halo shows that the Milky Way may once have been ball-shaped before it became a disc.
- Astronomers know there is a lot more matter in the halo than they can see. They know it is there, because they can measure the effect of its gravity. This invisible matter is called dark matter.
- No-one is certain what dark matter is.

GALAXY PROFILE

Shape	**Spiral**
Diameter of disc	**100,000 light years**
Average thickness of disc	**10,000 light years**
Diameter of central bulge	**12,000 light years**
Thickness of central bulge	**30,000 light years**
Number of stars	**200-400 billion**

A NASA photograph of the Milky Way.

AGE OF THE MILKY WAY

- Astronomers think the Milky Way formed soon after the Universe began.
- The Milky Way probably formed when the Universe was only about 200-300 million years old.
- This means the age of the Milky Way is about 13,000 to 13,600 million years.

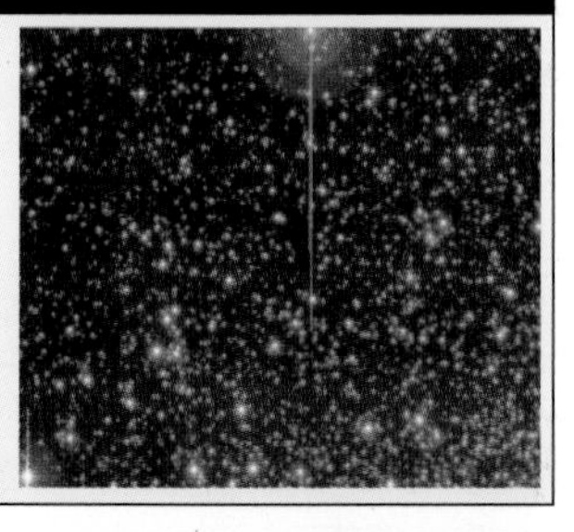

Globular cluster NGC 6397 contains some of the oldest stars in the Milky Way.

MILKY WAY CENTRE

- We cannot see the centre of the Milky Way because it is surrounded by thick clouds of gas and dust.
- Astronomers think there may be a giant black hole at the centre, containing as much matter as a million Suns.
- We are in no danger of falling into the Milky Way's black hole.
- The Sun is located about two-thirds of the way out from the centre of the Milky Way towards its edge.
- The centre is about 25,000 light years away from us.

An artist's impression of the Milky Way.

• See pages 48-49 for information on galaxies.

MANY ARMS

The Milky Way has four main spiral arms that curl out from the centre of the galaxy. These are the Norma arm; the Scutum-Crux arm; the Sagittarius arm and the Perseus arm. The Sun lies in a small arm called the Orion arm (also called the Local Arm).

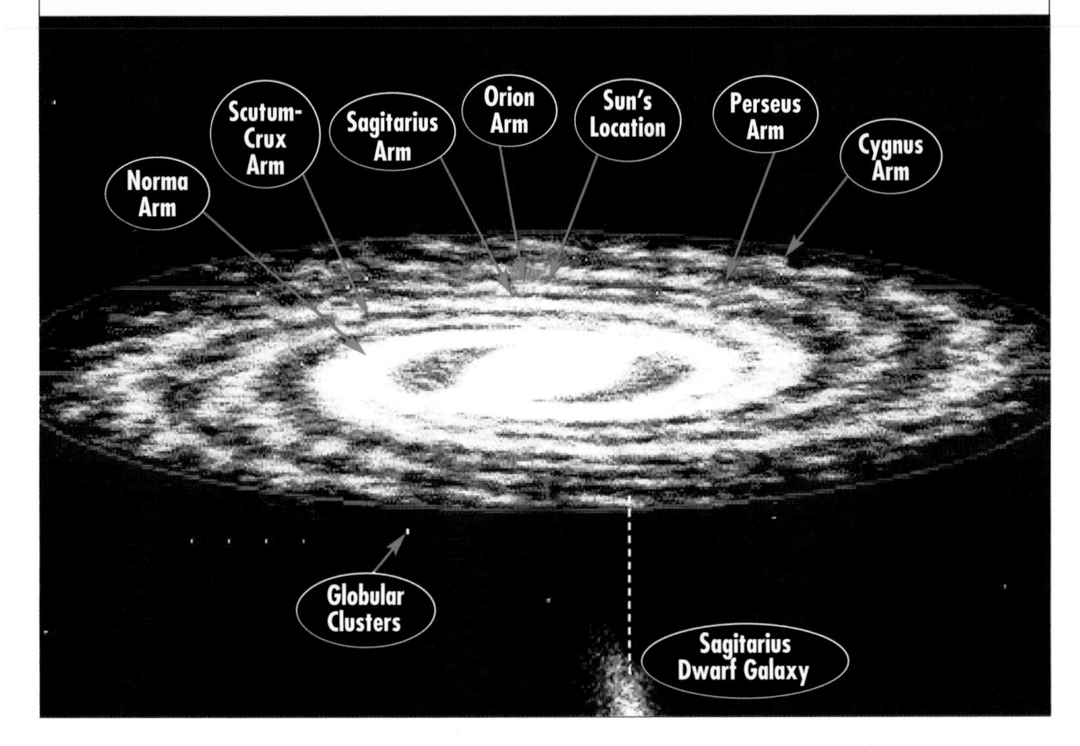

IN A SPIN

Sun's path around the Galaxy

- The Milky Way slowly spins as it moves through space like a giant cart-wheel.
- The Sun takes about 226 million years to complete one orbit.
- The Sun moves around the Milky Way at a speed of 220 km per second.
- The Sun has orbited the centre of the Milky Way about 25 times since its formation.

• See pages 12-13 for information on the Sun.

SPEEDING STARS

The majority of the stars in the Sun's vicinity in the Milky Way move around the Galaxy at around 30-50 km per second. There are, however, some stars that travel around twice as fast as that speed.

NEAREST GALAXIES

GALAXY	DISTANCE FROM EARTH
Canis Major Dwarf Galaxy	25,000 light years
Sagittarius Dwarf Elliptical Galaxy	81,000 light years
Large Magellanic Cloud	160,000 light years
Small Magellanic Cloud	190,000 light years
Ursa Minor Dwarf Galaxy	205,500 light years

The Sagittarius Dwarf Elliptical Galaxy.

FUTURE FATE

The Milky Way is slowly moving towards another galaxy called the Andromeda galaxy.

- Andromeda is bigger than the Milky Way.
- One day the two galaxies will collide, but they will not meet for several billion years.
- The two galaxies will probably merge and form a new galaxy.
- The new galaxy will not be a spiral like the Milky Way, but it will probably be elliptical in shape.
- The Milky Way has already swallowed up many small nearby galaxies.

The two illustrations below simulate what might happen when the Andromeda Galaxy hits ours. The central regions will merge into a single galaxy.

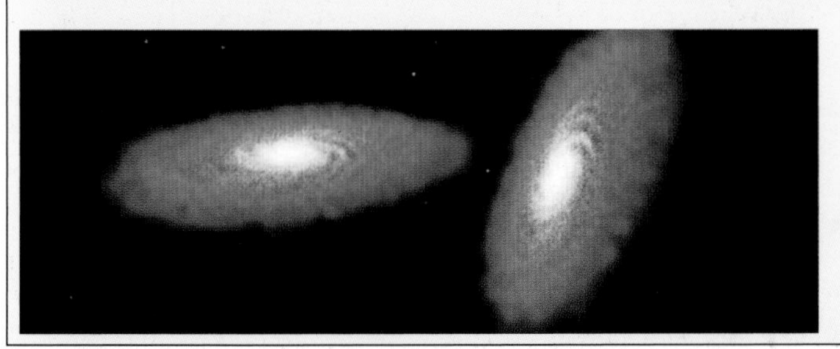

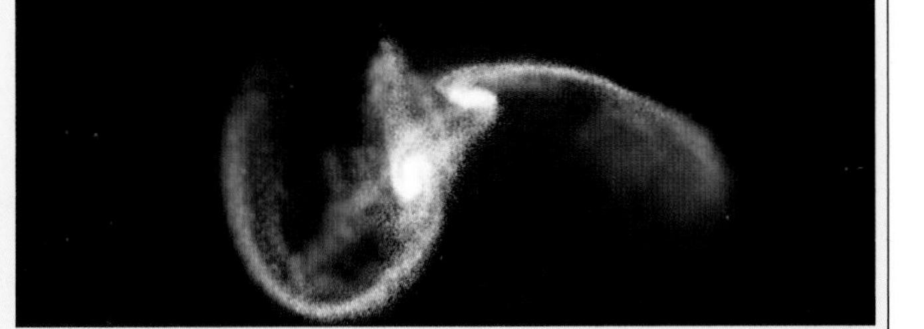

GALAXIES

There are billions of giant star groups like the Milky Way. They are called galaxies. Some galaxies are spiral in shape like the Milky Way. Others are different shapes. Galaxies are not spread evenly through space. They bunch together in groups, or clusters. The clusters themselves bunch together in bigger groups called superclusters. Our galaxy, the Milky Way, belongs to a cluster of about 30 galaxies called the Local Group. This is one of about 400 clusters of galaxies that form a grouping called the Local Supercluster.

GALAXY NAMES

- Galaxies known for a long time were named after their discoverer.
- The Large Magellanic Cloud is named after the explorer, Ferdinand Magellan, whose crew discovered it during the first voyage around the world in the 1520s.
- Some galaxies are named after the constellation they are found in.
- The Sagittarius galaxy is named after the Sagittarius constellation that it appears in.
- Today, galaxies are named differently. They are usually known by the name of a catalogue they are listed in and their number in the catalogue.
- The New Galactic Catalogue (NGC) lists hundreds of galaxies. They all have an NGC number.
- Some galaxies are also known by an 'M' number. They are galaxies that appear in a list drawn up in 1781 by the French astronomer Charles Messier. In this list, the Andromeda galaxy is M31.
- Some galaxies appear in more than one list and so they have more than one name or number. For example, NGC 598 is also known as M33.

GALAXY SHAPES

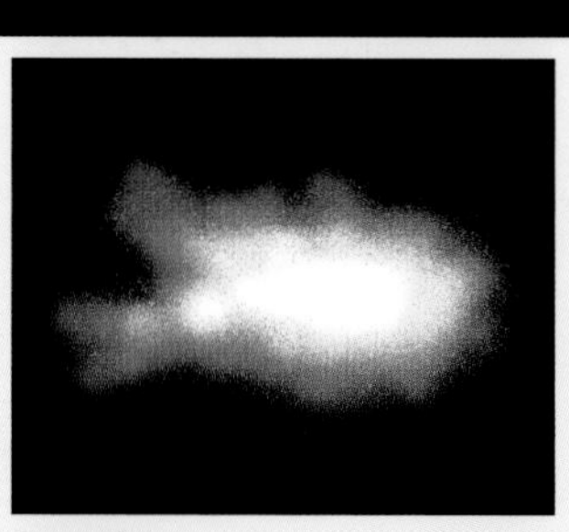

Irregular

Elliptical

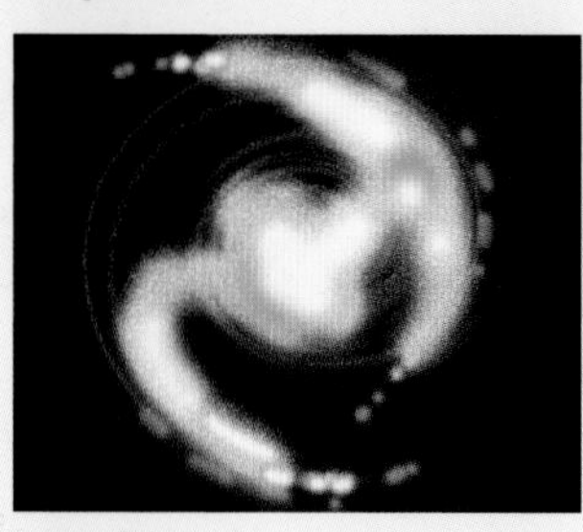

Barred Spiral

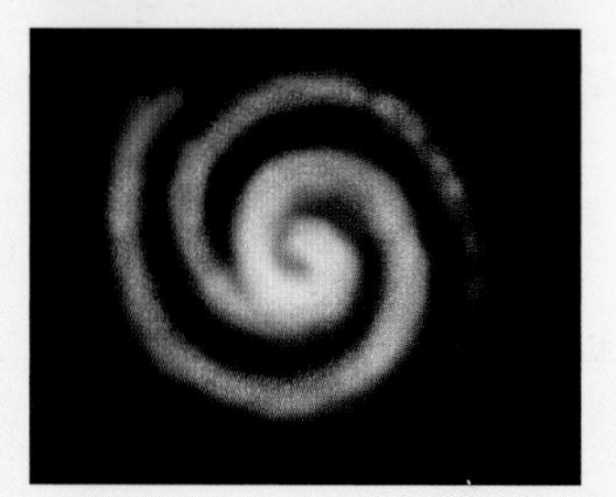

Spiral

All galaxies have one of three basic shapes – spiral, elliptical or irregular.

- Nearly all galaxies are either spiral or elliptical in shape.
- Elliptical galaxies may be perfectly ball-shaped or they may be flattened, stretched oval shapes.
- Spiral galaxies usually have a central bulge of older stars surrounded by a disc of matter where new stars are forming.
- Stars in a spiral galaxy's disc trace out the shapes of spiral arms.
- About half of spiral galaxies have a straight bar of stars through the central bulge.
- These galaxies are called barred spirals.
- Irregular galaxies have no definite shape.
- Some irregular galaxies look asif they were once spirals. They may have changed shape when they collided with other galaxies.
- Other irregular galaxies are confused jumbles of stars.
- This way of dividing galaxies into groups according to their shape was devised by the US astronomer Edwin Hubble in the 1920s.

• See page 11 for more information on Edwin Hubble.

ACTIVE GALAXIES

Active galaxies are galaxies that give out far more light and other energy than their stars should produce.

- In the 1960s astronomers discovered objects that looked like stars, but they were distant galaxies with a brilliant core 100 times brighter than a normal galaxy.
- These active galaxies are called quasi-stellar objects, or quasars.
- The blindingly bright core of a quasar is only about the size of our Solar System, but brighter than scores of Milky Way galaxies.
- The huge amount of light pouring out of a quasar seems to be coming from matter falling into a black hole.
- The black hole at the centre of a quasar is massive, perhaps a billion times the mass of the Sun.

The Chandra X-Ray observatory (model shown here) will allow researchers to obtain better X-ray images of quasars.

HOW GALAXIES MOVE

Astronomers can tell a lot about a distant galaxy by studying the light and other energy that it gives out.

- A galaxy gives out known patterns of light according to the chemical elements it contains.
- When astronomers analyse light from distant galaxies, these distinctive patterns look wrong. The colours seem to have been shifted towards the red end of the rainbow-spectrum of visible light. This is called red shift.
- Red shift happens because the light waves have been stretched. Stretching light waves changes the colours they make.
- Light waves are stretched like this when they come from something that is rushing away from us.
- This indicates that all the galaxies we can see are flying away from us.

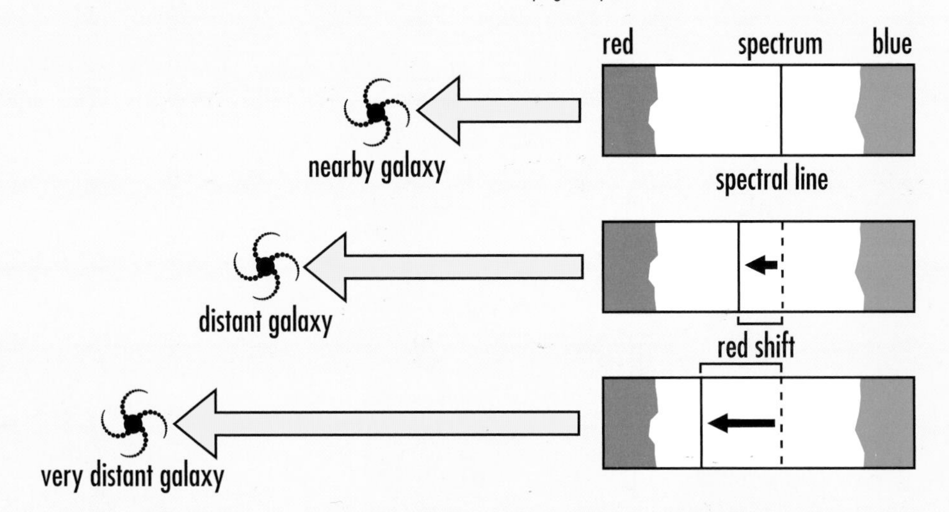

HOW MANY GALAXIES?

- No-one knows the exact number of galaxies that exist today.
- Astronomers see galaxies wherever they look in the sky.
- When they point their most powerful telescopes and most sensitive cameras at a tiny part of the sky only one tenth the diameter of the Moon that seemed to be empty, astronomers can see 10,000 galaxies there.
- Astronomers believe there are billions of galaxies in total.

• See pages 44-45 for information on telescopes.

FARTHEST GALAXIES

- The furthest galaxies observed so far are about 13 billion light years away from us.
- These galaxies are so far away that light from them has travelled across the Universe for 13 billion years to reach us.
- Astronomers are seeing these distant galaxies as they looked 13 billion years ago when the light left them.
- This was only just after the Universe formed, so even if there are farther galaxies, light from them would not have time to reach us. These distant galaxies may not be there any more. Even if they still exist, they probably look very different today.

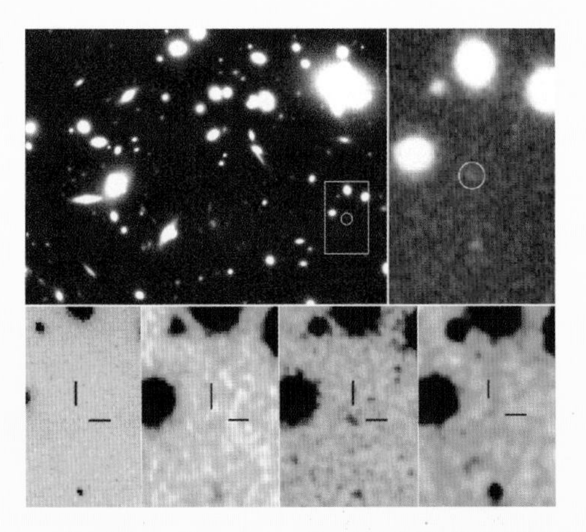

European Southern observatory image of Abell 1835 IR1916, the most distant galaxy yet discovered, located about 13,230 million light-years away.

LARGEST LOCAL GALAXIES

Galaxy	Type	Diameter	Distance
Andromeda (below)	Spiral	130,000 light years	2.5 million light years
Milky Way	Spiral	100,000 light years	-
NGC 598	Spiral	60,000 light years	2.7 million light years
Large Magellanic Cloud	Irregular	30,000 light years	160,000 light years
Small Magellanic Cloud	Irregular	25,000 light years	180,000 light years

UNIVERSE

The Universe is all the matter, energy, space and time that exists everywhere. Scientists have tried many theories to explain how and when the Universe began. Today, most scientists agree that the Universe burst into existence about 13.7 billion years ago in a huge explosion called the Big Bang. It flung matter and energy in all directions and produced the Universe we see today.

DARK MATTER

- Scientists believe that there is a lot more matter in the Universe than we can see.
- Most of the matter in the Universe is invisible, but scientists can detect the effect of its gravity, which is how we know it exists.
- This invisible matter is called Dark Matter.
- No-one yet knows what Dark Matter is made of.

EXPANSION OF THE UNIVERSE

The galaxies all appear to be shooting away from us at very great speeds.

This is not because we occupy any special position in the Universe – exactly the same thing would be observed from any other galaxy. In fact, it is not the galaxies that are moving, but the space between them that is expanding. Imagine sticking stars on a balloon and then blowing the balloon up. The stars would seem to move apart as the balloon inflates. The expansion of the Universe is a process that has been occurring over billions of years.

COSMOLOGY

- The branch of science concerned with studying the origin and development of the Universe is called cosmology.
- Scientists who study cosmology are called cosmologists.
- Cosmologists may never look through telescopes. They often rely on information about stars, galaxies and other objects from the people who actually observe them – astronomers.

The cosmologist Albert Einstein.

• See pages 8-11 for information on astronomers.

NEW THEORIES

Scientists are still developing new theories to explain the origin and evolution of the Universe.

- According to string theory, particles of matter are actually tiny vibrating loops and strands of energy.
- The strings may lie on sheets or tubes of space called membranes, or branes.
- String theory predicts that there are another six dimensions as well as the three dimensions in space and one in time that we already know about.
- Another prediction of string theory is that there may be many more Universes as well as ours.
- These parallel Universes are called a Multiverse.

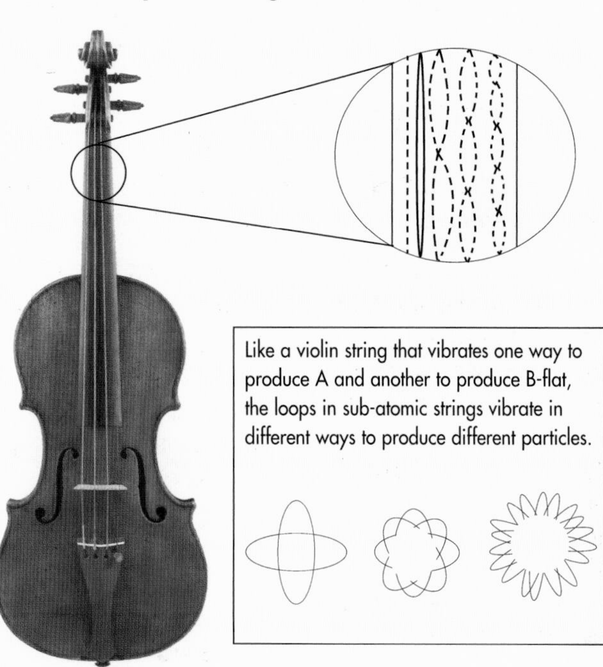

Like a violin string that vibrates one way to produce A and another to produce B-flat, the loops in sub-atomic strings vibrate in different ways to produce different particles.

THE BIG BANG

Most scientists think the Universe began from a tiny speck which exploded at the start of space and time, called the Big Bang.

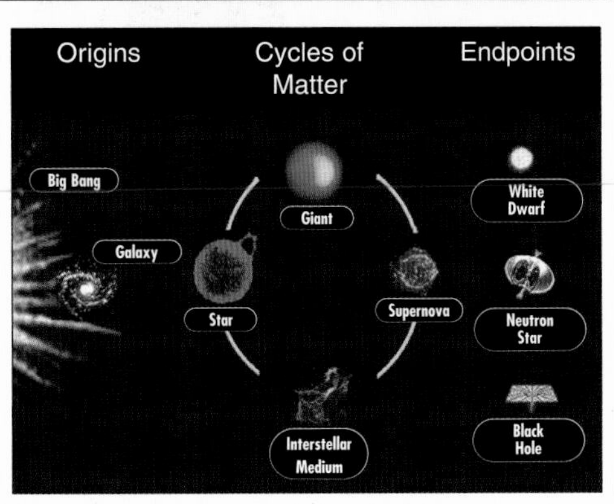

- In its first moment of existence, the Universe was unimaginably hot.
- It expanded rapidly during a stage called inflation.
- As the Universe expanded, it cooled.
- Within a fraction of a second, the first particles were produced.
- One second after the Big Bang, the temperature had fallen to 10 billion degrees.
- After three minutes, the temperature was one billion degrees – cool enough for particles of matter to join together.
- About 300,000 years after the explosion, it was cool enough for whole atoms to form.
- Hydrogen, helium and other simple elements formed.
- Matter was not spread evenly through the early Universe.
- Denser clumps of matter developed into the first stars and galaxies only about 200 million years after the big Bang.
- The Big Bang produced an expanding Universe, which we see today.
- The Big Bang also led to the creation of hydrogen and helium in the proportions we see today.
- In the 1950s, scientists suggested that if the Big Bang really happened, its echo should still exist.
- As the Universe expanded, the super-hot radiation that filled it spread out and cooled.
- This radiation still exists today.
- It is called the cosmic background radiation or cosmic microwave background.
- Two astronomers, Arno Penzias and Robert Wilson, detected the Cosmic Background Radiation in 1965.
- In 1992, the Cosmic Background Explorer (COBE) satellite made a map of the background radiation.

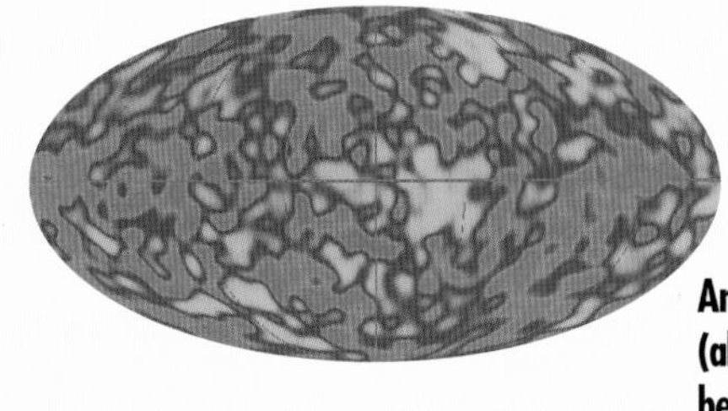

An image of the Universe taken by the COBE satellite (above right) confirming a temperature of 270 degrees below zero.

COSMOLOGISTS TIMELINE

Ptolemy (87-150 AD)
His book *Almagest* stated Earth is at the centre of the Universe.

Nicolaus Copernicus (1473-1543)
Proposed that the Sun is at the centre of the Universe.

Tycho Brahe (1546-1601)
Said that the stars are not carried around on crystal spheres, but fly through the sky unsupported.

Galileo Galilei (1564-1642)
Made observations showing that Copernicus was right.

Sir Isaac Newton (1642-1727)
Developed the theory of gravity that extended far beyond Earth to the Moon and space.

Friedrich Bessel (1784-1846)
Measured the distances to some of the closest stars.

William Parsons (1800-1867)
Made the first detailed observations of spiral galaxies.

Albert Einstein (1879-1955)
Showed the relationship between space and time and matter and energy.

Harlow Shapley (1885-1972)
Calculated the size of the Milky Way galaxy and found the location of the Sun within it.

Edwin Hubble (1889-1953)
Discovered that other galaxies are outside the Milky Way and that the Universe is expanding.

THE FUTURE OF THE UNIVERSE

What will be the eventual fate of the Universe? Everything depends on how dense the Universe is. There are three possibilities;

- If the density is higher than a certain value (the 'critical density'), then the Universe will eventually stop expanding and collapse in on itself. This is called the Big Crunch Scenario.
- If the density is less than critical, then the Universe will just go on expanding, and the temperature of everything in the Universe will plummet. Everything will become freezing cold. This scenario is called the Heat Death Scenario.
- Finally, if the density is just borderline, the Universe will expand less and less but will not collapse. This is called the Flat Universe Scenario.

A balloon experiment called Boomerang was sent high into the atmosphere above the Antarctic, to measure the 'bumpiness' in the cosmic background. This measurement shows that the Universe is actually flat. In other words, neither the Big Crunch nor the Heat Death scenario will happen.

The balloon experiment Boomerang with a projection of its results superimposed behind.

HUMANS IN SPACE

People have been travelling into space since 1961. They flew in a variety of different spacecraft. The first manned spacecraft were just big enough to fit one person inside. Later spacecraft could fly three people around the world or take them all the way to the Moon and back. At first all of these spacecraft were used for only one space mission each. The US Space Shuttle is different. It can be flown again and again. Until 2003, Russia and the USA were the only countries with a manned space programme, but in that year China launched its first manned space mission. Shenzhou-5 carried Yang Liwei into space on October 15th. He returned home safely after orbiting Earth 14 times.

APOLLO CREWS

MISSION	CREW
Apollo 7	Schirra, Eisele, Cunningham
Apollo 8	Borman, Lovell and Anders
Apollo 9	McDivitt, Scott and Schweickart
Apollo 10	Stafford, Cernan and Young
Apollo 11	Armstrong, Aldrin and Collins
Apollo 12	Conrad, Gordon and Bean
Apollo 13	Lovell Jr, Swigert Jr and Haise Jr
Apollo 14	Shepard Jr, Roosa and Mitchell
Apollo 15	Scott, Worden and Irwin
Apollo 16	Young, Mattingly and Duke Jr
Apollo 17	Cernan, Evans and Schmitt

VOSTOK (USSR/RUSSIA)

Vostok was the first manned spacecraft and the culmination of a space race between the United States and the Soviet Union.

- The Vostok capsule was a small sphere only 2.5 metres across.
- The Soviet Union called its space travellers cosmonauts.
- Each capsule carried one cosmonaut into space for missions lasting up to five days.

The Vostok 1 space capsule.

- The first cosmonaut was Yuri Gagarin.
- He made the first ever manned spaceflight on 12 April 1961.
- On the way back Gagarin was ejected from his capsule at a height of 7,000 metres and landed by parachute.
- The first manned spaceflight lasted less than two hours from launch to landing.
- Vostok 6 launched Valentina Tereshkova, the first woman in space, on 16 June 1963.

The cosmonaut Yuri Gagarin.

GEMINI (USA)

Gemini 4 astronaut Ed White during his spacewalk over El Paso, Texas.

- Gemini was a two-person spacecraft, launched by the Titan rocket. There were 10 manned Gemini spaceflights between 1965 and 1966.
- Ed White made the first spacewalk by an American astronaut during the Gemini 4 mission.
- Gemini allowed astronauts to practice all the manoeuvres that would be needed for a moon-landing mission.
- The longest Gemini mission was Gemini 7, which lasted for 13 days 18 hours.

MERCURY (USA)

The USA answered Vostok with its own one-person space capsule, called Mercury.

- After re-entry Mercury capsules landed in the sea.
- America's first astronaut was Alan Shepard. He made a 15-minute sub-orbital flight (into space but not around the Earth) in Mercury 3 on 5 May 5 1961.
- Alan Shepard was the only Mercury astronaut who would later walk on the Moon.

The recovered Mercury space capsule.

- Virgil Grissom's Mercury 4 capsule sank soon after it splashed down. Grissom was rescued.
- Mercury 6 carried the first American to go into orbit, John Glenn, on 20 February 1962.
- There was an emergency during Glenn's flight when ground controllers thought his spacecraft's heat shield had come loose, but he landed safely.
- There were 6 manned Mercury missions between 1961 and 1963.

APOLLO (USA)

The Apollo spacecraft were designed to take three astronauts to the Moon and land two of them on the Moon.

- The Apollo spacecraft were made of three modules – the Command Module, the Service Module and the Lunar Excursion Module (LEM).
- The crew spent most of their time in the Command Module, supplied with air, electricity and rocket power by the Service Module.
- The LEM was the part of the spacecraft that landed on the Moon.
- The tiny Command Module was the only part of the spacecraft to return to Earth and splash down in the Pacific Ocean.
- The spacecraft was launched by the giant Saturn V rocket.
- The crew of Apollo 8 were the first humans to orbit the Moon.
- After four manned test-flights, Apollo 11 landed Neil Armstrong and Buzz Aldrin on the Moon's surface in 1969, while Michael Collins circled the Moon.
- The Apollo 13 spacecraft suffered a serious explosion on its way to the Moon. The crew returned to Earth safely.
- The Apollo 13 crew travelled further from Earth than anyone else – 400,171km.
- The last three Apollo missions carried an electric car called the Lunar Rover, used to travel across the Moon's surface.
- The last Apollo Moon-landing mission was Apollo 17 in 1972. The Apollo 17 crew stayed on the Moon for the longest time – 75 hours.
- Over six missions 12 Apollo astronauts landed on the Moon and brought Moon rocks back to Earth. They also left packages of scientific instruments on the Moon's surface.
- Left-over Apollo hardware was used for a joint US-Soviet mission called the Apollo-Soyuz Test Project, and also for the Skylab space station.

• See pages 20-21 for information on the Moon.

SOYUZ (USSR/RUSSIA)

Soyuz was initially designed as part of the USSR's Moon mission, and is the longest serving manned spacecraft in the world.

- A Soyuz spacecraft has three modules – an instrument module, a re-entry module and an orbital module. The spacecraft is 7.0 metres long and 2.7 metres in diameter.
- The re-entry module lands by parachute on the ground. It fires rockets just before touch-down to cushion the impact.
- The first manned Soyuz mission took place in 1967. Unfortunately, the spacecraft crashed and killed the pilot, Vladimir Komarov. Since then, however, Soyuz has been a highly successful and reliable spacecraft.
- Soyuz spacecraft ferry cosmonauts and astronauts to and from the new International Space Station.
- A type of unmanned Soyuz spacecraft called Progress was developed to deliver supplies to space stations.

SPACE SHUTTLE (USA)

The development of a reusable space launcher was the next stage in man's exploration of space.

- The Space Shuttle Orbiter is the size of a small airliner. It is 37 metres long with a wingspan of 24 metres.
- A payload bay carries satellites, laboratories and scientific instruments into space.
- The Space Shuttle has a crew of two pilots and up to 5 other astronauts.
- It is launched with the help of two solid rocket boosters and an external fuel tank, which supplies fuel to the three big engines in the Orbiter's tail.
- The first Space Shuttle was launched in 1981, crewed by John Young and Bob Crippen.
- Since 1981, there have been more than 100 Space Shuttle missions.
- The oldest Shuttle astronaut was John Glenn, at the age of 77 years. He was the same John Glenn who made the first US orbital flight in 1962.
- There are currently three Orbiters – Discovery, Atlantis and Endeavour.
- Two other Orbiters, Columbia and Challenger, were lost along with their crews in accidents.

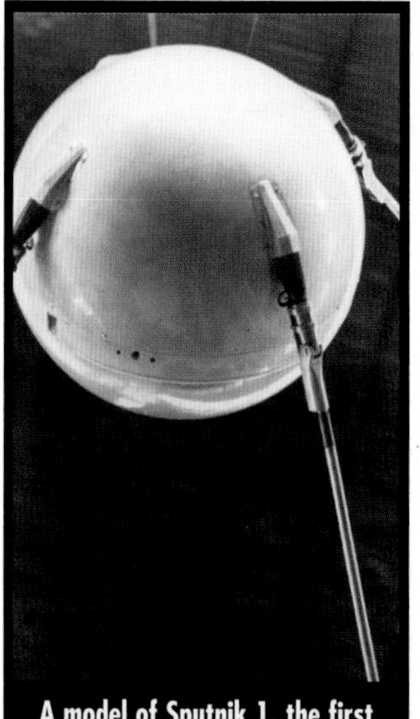
A model of Sputnik 1, the first space satellite.

SPACE PROBES

People have travelled only as far as the Moon, but unmanned space probes have toured almost the entire Solar System. The first probes were blasted into space towards the end of the 1950s when the Cold War between the United States and the USSR was at its height. They have explored the surface of Mars, created maps of Venus, taken close-up photographs of nearly all the planets and many of their moons, orbited Saturn and even left the Solar System altogether. There are too many space probes to list in full, but details of a few of the most important probes are given here.

SPUTNIK (USSR/RUSSIA)

PROBE	LAUNCHED	DETAILS
Sputnik 1	4 Oct 1957	The world's first artificial satellite
Sputnik 2	3 Nov 1957	Carried dog Laika, the first living space traveller
Sputnik 3	15 May 1958	Studied Earth's atmosphere and solar radiation
Sputnik 4	15 May 1960	Unmanned test-flight for Vostok 1
Sputnik 5	19 Aug 1960	Carried two dogs, Belka and Strelka
Sputnik 6	1 Dec 1960	The second test-flight for Vostok. Recovery failed.
Sputnik 7	4 Feb 1961	A test-flight for Venus probe Sputnik 8
Sputnik 8	12 Feb 1961	Launched Russia's first Venus probe
Sputnik 9	9 Mar 1961	Carried a dog called Chernushka
Sputnik 10	25 Mar 1961	Carried a dog called Zvezdochka

Laika, the first animal in space.

PIONEER (USA)

The Pioneer Venus Orbiter, launched in May 1978.

PROBE	LAUNCHED	DETAILS
Pioneer 1	11 Oct 1958	Mapped the Van Allen radiation belts around Earth
Pioneer 2	8 Nov 1958	Intended to reach the Moon, but failed
Pioneer 3	6 Dec 1958	Failed to reach the Moon
Pioneer 4	3 Mar 1959	Passed within 60,000 km of the Moon
Pioneer 5	11 Mar 1960	Entered solar orbit, sent solar flare and solar wind data
Pioneer 6	16 Dec 1965	Entered solar orbit and studied the Sun's atmosphere
Pioneer 7	17 Aug 1966	Combined with Pioneer 6 to study the Sun
Pioneer 8	13 Dec 1967	Joined Pioneers 6 and 7 to study the Sun
Pioneer 9	8 Nov 1968	Joined Pioneers 6, 7 and 8 to study the Sun
Pioneer 10	3 Mar 1972	Passed within about 130,000 km of Jupiter's cloud-tops
Pioneer 11	5 Apr 1973	Photographed Jupiter's south pole
Pioneer Venus Orbiter	20 May 1978	Orbited Venus, studying its atmosphere
Pioneer Venus Multiprobe	8 Aug 1978	Dropped four probes into Venus's atmosphere before entering the atmosphere and burning up

VENUS PROBES (USSR/RUSSIA)

PROBE	LAUNCHED	DETAILS
Venera 1	12 Feb 1961	First interplanetary flight passed within 100,000 km of Venus
Venera 2	12 Nov 1965	Passed within 24,000 km of Venus
Venera 3	16 Nov 1965	First spacecraft to land on another planet
Venera 4	12 Jun 1967	Transmitted data during descent through atmosphere
Venera 5	5 Jan 1969	Transmitted data during descent through atmosphere
Venera 6	10 Jan 1969	Transmitted data during descent through atmosphere
Venera 7	17 Aug 1970	Transmitted data from the surface of Venus
Venera 8	27 Mar 1972	Transmitted data from the surface of Venus
Venera 9	8 Jun 1975	First spacecraft to send pictures from the surface of another planet
Venera 10	14 Jun 1975	Transmitted pictures from the surface of Venus
Venera 11	9 Sep 1978	Fly-by probe dropped a lander onto Venus
Venera 12	14 Sep 1978	Fly-by probe dropped a lander onto Venus
Venera 13	30 Oct 1981	Landed using radar maps made by Pioneer-Venus spacecraft
Venera 14	4 Nov 1981	Fly-by probe dropped a lander onto Venus
Venera 15	2 Jun 1983	Orbited Venus and mapped surface by radar
Venera 16	6 Jun 1983	Orbited Venus and mapped its surface by radar

Venera 13 replica at the Cosmos Pavillion in Moscow.

• See pages 16-17 for information on Venus.

VIKINGS TO MARS (US)

In 1976, the US landed two Viking spacecraft on the red planet.

PROBE	LAUNCHED
Viking 1	20 Aug 1975
Viking 2	9 Sep 1975

- After the US Mariner Mars probes in the 1960s and 1970s had revealed Mars to be a cratered Moon-like world, scientists were keen to land a spacecraft on the surface.
- Both of the Viking missions placed an orbiter in orbit around Mars,while a lander descended to the surface.
- The landers tested the Martian soil and atmosphere, and sent data back to Earth via the orbiters.
- Sadly, they found no definite signs of life.
- The Vikings also took the first high-quality close-up colour photographs of the Martian surface (see below).

• See pages 22-23 for information on Mars.

RECENT PLANETERY PROBES

PROBE	LAUNCHED	DETAILS
Magellan	1989	Orbited Venus and produced detailed maps of its surface by radar
Galileo	1989	Surveyed Jupiter and its moons before plunging into Jupiter's atmosphere in 2003
Mars Pathfinder	1996	Landed the rover Sojourner on Mars in 1997
Cassini-Huygens	1997	Touring Saturn and its moons, landed the Huygens mini-probe on Saturn's moon, Titan, in 2005
Mars Exploration	2003	The rovers Spirit and Opportunity landed on Mars in 2004 and found evidence of water.

10 EARLY MOON PROBES

The Moon has been visited by a series of probes, some manned and some robotic.

In 1959, Luna 2 became the first space probe to get to the Moon, while its successor took the first photographs of the Moon's far side. In 1966 Luna 9 made the first controlled landing on the Moon, and in 1967 Surveyor 3 (pictured right) dug a trench in the Moon's soil. Luna 16 brought back a sample of Moon dust to Earth in 1970 while the same year its successor landed a rover vehicle, Lunokhod 1.

SALYUT SPACE STATIONS		
STATION	LAUNCHED	RETURNED
Salyut 1	19/04/71	11/10/71
Salyut 2	04/04/73	28/05/73
Salyut 3	25/06/74	24/08/75
Salyut 4	26/12/74	03/02/77
Salyut 5	22/06/76	08/08/77
Salyut 6	29/09/77	29/07/82
Salyut 7	19/04/82	07/02/91

LIVING IN SPACE

People can stay in space for months or years at a time by living on orbiting space stations. Russia has launched eight space stations – seven Salyut stations and Mir. Meanwhile, the USA used rockets and spacecraft left over from the Apollo Moon missions to create the Skylab space station. Now Russia, the USA and more than a dozen other countries have come together to built the ISS, the International Space Station ('Freedom'). Space stations allow scientists to carry out long-term experiments and observations, and to study how people adapt to long periods in space.

SALYUT SPACE STATIONS

Salyut 1 was launched by the Soviet Union in 1971. It was the world's first space station.

- The Salyut space stations each measured 13 metres long by 4.2 metres across.
- Salyuts 1-5 had one docking port, where a transfer spacecraft could attach.
- Salyuts 6 and 7 had two docking ports.
- A Salyut space station weighed about 19 tonnes
- Salyut 6 and 7 could be re-fuelled, which enabled them to stay in orbit for longer.
- Salyut 7 spent nearly nine years in orbit

The 13-metre long Salyut 7 on display in 1984 in Helsinki, Finland.

SKYLAB MISSIONS		
MISSION	LAUNCHED	LENGTH
Skylab 1	14/05/73	N/A
Skylab 2	25/05/73	28 days
Skylab 3	28/07/73	59 days
Skylab 4	16/11/73	84 days

Skylab astronaut Alan Bean.

SKYLAB SPACE STATION

Skylab was the United States' only space station.

- Launched on May 14, 1973, it spent more than six years in orbit, but Skylab was shaken so much during launch that one of its solar panels was torn off.
- Skylab was 36 metres long, with a spacecraft docked, and up to 6.6 metres wide.
- Three three-man Skylab crews took tens of thousands of photographs of the Earth and studied the Sun using Skylab's solar observatory.
- Skylab re-entered the Earth's atmosphere on July 11, 1979.

Skylab in orbit at the end of its mission.

MIR SPACE STATION

The space station Mir with the shuttle Atlantis.

In 1986 the Soviet Union launched the first 20 modules of a new space station called Mir ('peace').

- Five more modules were added until the mid-1990s (see below).
- Mir spent just over 15 years in orbit.
- The US Space Shuttle was able to dock with Mir.
- Seven Shuttle astronauts spent a total of 32 months on Mir.
- In 1997, an unmanned Progress supply craft crashed into Mir, damaging its hull and solar panels.
- The longest time spent in space at one time is 437 days, achieved by Russian cosmonaut Valeriy Polyakov on Mir from 8 January 1994 to 22 March 1995.
- Towards the end Mir had many breakdowns and mould was growing on its equipment.
- Mir re-entered the Earth's atmosphere on 23 March 2001.
- Several large pieces of the 130-tonne space station survived re-entry and splashed down in the South Pacific Ocean, east of New Zealand.

MIR MODULES

MODULE	LAUNCHED	USED FOR
Core module	1986	Command centre and accommodation
Kvant-1	1987	Astronomical observatory
Kvant-2	1989	Airlock for spacewalks
Kristall	1990	Processing materials
Spektr	1995	Remote sensing
Piroda	1996	Remote sensing

THE INTERNATIONAL SPACE STATION

ISS - HISTORY

In 1998, the first two modules of the International Space Station were launched and joined together in orbit.

- The first parts were called the Zarya module. Zarya was launched by a Russian Proton rocket in November 1998.
- In December 1998, the Space Shuttle added the Unity Node connecting segment.
- In July 2000, the Zvezda service module was added.
- The first crew, called Expedition 1, arrived in November 2000.
- The Destiny module was added in February 2001.
- Airlocks, docking ports for spacecraft and connecting frames have been added too.

• See pages 50-51 for information on the Space Shuttle.

The International Space Station.

ISS - FUTURE

- Its solar panels will have a span of 73 metres, almost the length of a soccer pitch.
- The living space inside will be bigger than in a Jumbo Jet.
- It will be about four times the size of Mir or five times the size of Skylab.
- It will have six laboratories, two habitation modules and two logistics modules.
- The ISS orbits at a height of 400 kilometres above the surface.
- NASA is building a human-like robot, called Robonaut, to help ISS astronauts.
- Shuttle problems from 2003 have delayed the probabe completion of the main ISS units until at least 2010.

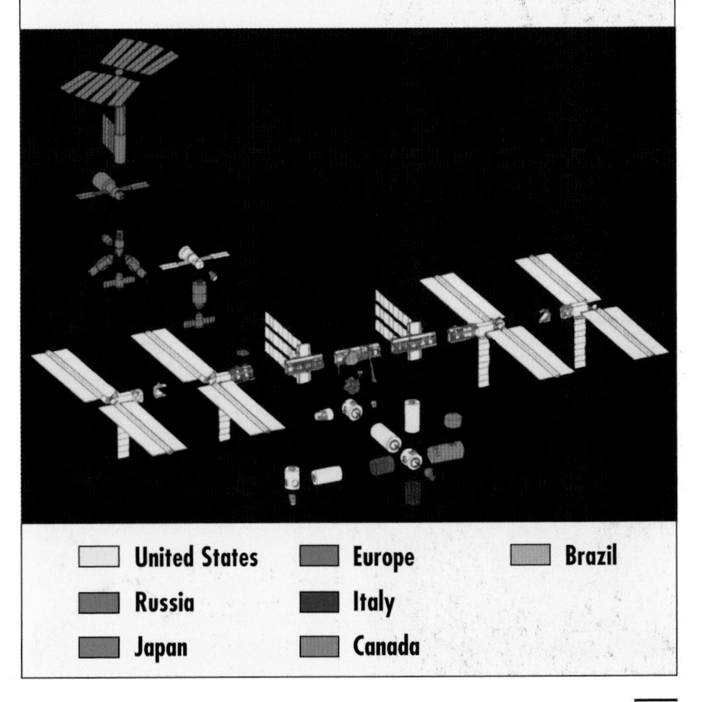

United States | Europe | Brazil
Russia | Italy
Japan | Canada

GLOSSARY

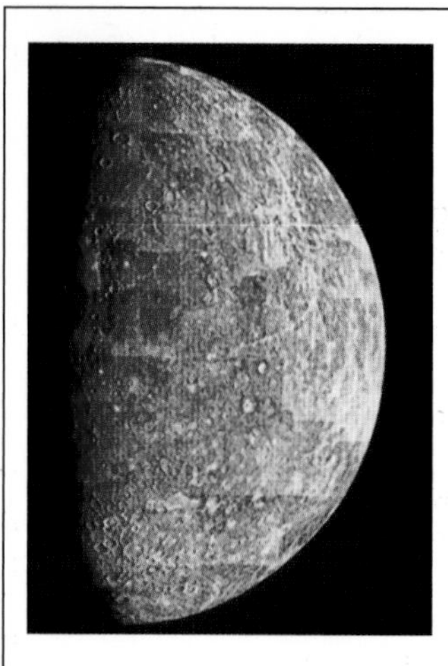

Arachnoid A type of volcano with ridges around it which resembles a spider's body and legs when seen from above. First discovered and named on planet Venus.

Asteroid A lump of rock or metal, usually smaller than a planet and without an atmosphere or moons orbiting it, but also larger than a meteoroid. Most asteroids orbit the Sun. They are sometimes called 'minor planets'.

Asteroid Belt
The zone between the outermost inner planet Mars, and the innermost outer planet Jupiter, where millions of asteroids orbit the Sun.

Astronomy
The study of planets, stars, moons and other space bodies, also of space itself and the whole of the Universe. Astronomy is partly a practical science which involves observing and cataloguing – *see also Cosmology*.

Astronomer
Someone who studies the planets, stars, moons and other bodies and objects in space.

Atmosphere The layer of gases surrounding a planet or star.

Axis An imaginary line passing through the middle of a star or planet, around which it spins or rotates.

Big Bang An unimaginably gigantic explosion, thought to have happened at the beginning of the Universe, more than 13 billion years ago, perhaps when space, time and matter began.

Black hole A very small, dense, dark area of space with immensely powerful gravity, created when a star collapses to 'less than nothing' and pulls in everything around it, including light.

Coma A bright glowing cloud or 'halo' around a body such as a comet.

Comet A relatively small ball of rock and ice, a 'dirty snowball' orbiting the Sun on a very lop-sided path that may take it far beyond the distance of Pluto, to the Oort Cloud.

Constellation A pattern or picture seen in a group or cluster of stars as they are viewed from Earth.

Core The central part of a planet, moon, star or other space object.

Corona The glowing ring or 'halo' around the Sun, best seen during a solar eclipse when the Moon passes in front of the Sun.

Cosmology The development of astronomy dealing with the origin and evolution of the Universe as a whole, and how its parts work.

Crater A bowl or dish-shaped hollow on a planet, moon or asteroid, caused by another object crashing into it.

Crust The hard, rocky outer layer of a planet such as the Earth.

Dark matter Invisible material that we cannot detect using scientific methods, but which is thought to make up a large percentage of the Universe.

Day The amount of time it takes a planet or moon to spin round once on its axis, so that its nearest star returns to the same point in its sky.

Dwarf star A star that is smaller than the Sun.

Eclipse When one space object, such as a moon, goes between another object and a star, as when the Moon moves between the Sun and Earth, casting a shadow on the Earth.

Equator An imaginary line around the middle of a planet or moon, at right angles to its axis of rotation (spin).

Galaxy A huge grouping or cluster of stars, planets and other objects, held together in space by gravity, and with immense distances of almost empty space to the next galaxies.

Gas giants Jupiter, Saturn, Uranus and Neptune, the four largest outer planets in the Solar System, which are made mostly of gases.

Giant star A star that is bigger than the Sun.

Gravity A force that makes any object or matter pull or attract other objects towards it. 'Gravity' often refers to Earth's gravity, and 'gravitational force' to the general name of the force which acts everywhere throughout the Universe. The gravitational force is one of only four fundamental forces in the Universe (the others being the electromagnetic force, the strong nuclear force and the weak nuclear force).

Hemisphere Half of a star, planet, moon or similar object, usually either above (north) or below (south) of its equator.

Inner planets Mercury, Venus, Earth and Mars, the four smaller and mainly rocky planets of the inner Solar System, closest to the Sun.

Kuiper Belt A zone of orbiting asteroid or comet-like objects in the outer Solar System, orbiting the Sun beyond Neptune. Its existence was first proposed in 1951 by US astronomer Gerard Kuiper.

Lander A spacecraft, or part of one, designed to land on another space object like a planet or moon, or to plunge into the atmosphere of a gassy object like Jupiter. It can be crewed by astronauts, or unmanned and remote-controlled.

Lens A curved piece of glass that bends or refracts light, as used in telescopes, binoculars and microscopes.

Lunar To do with the Moon (Earth's moon).

Lunar eclipse When the Earth passes between the Sun and the Moon, casting a shadow on the latter.

Mantle A layer of rock or other materials that lies between the core and the outer surface of a planet or moon.

Mass The amount of matter in an object – that is, the numbers and types of atoms or their subatomic particles, independent of any gravity acting on them.

Meteor A meteoroid that enters Earth's atmosphere and burns up, appearing as a bright streak of light also called a 'shooting star' or 'falling star'.

Meteorite A meteoroid that falls all the way to the Earth's surface.

Meteoroid A small chunk of rock, metal, ice or a mixture of these, usually broken off a comet or asteroid.

Milky Way Our 'home' galaxy which contains the Solar System.

Moon A space object that orbits a planet.

NASA (National Aeronautics and Space Administration) The government space agency of the USA.

Nebula A huge cloud of gas and dust in space, where new stars are often forming. A planetary nebula is a 'shroud' of gas thrown off by an overheating or fading star.

Neutron star The small, incredibly dense or 'heavy' remnant of a star which has exploded and collapsed into a ball of subatomic particles called neutrons.

Observatory A building or site where telescopes are used to observe (watch) objects in space.

Orbit To go round and round another object. The orbit of a planet or moon is its path around another object, such as the Sun. Most orbits are not circular but elliptical or 'oval'.

Orbiter A spacecraft or part of one that orbits its destination, rather than carrying out a fly-by or trying to land on the surface.

Oort Cloud A huge, ball-shaped cloud of comets and similar objects surrounding the whole Solar System.

Planet A large, spherical, rocky and/or gassy object orbiting a star. Planets can be defined and distinguished from other objects such as asteroids by their size, whether they orbit around a star, whether they have an atmosphere, and other features.

Pole The point on a rotating space object where the imaginary line around which it spins (the axis of rotation) passes through its surface.

Pulsar A rapidly rotating neutron star, that sends out 'beams' of radio and other energy as it spins, like a lighthouse, and so appears to flash or blink on and off.

Radar Radio Direction And Ranging – a way of measuring shapes and distances of objects by bouncing or reflecting radio waves off them, and detecting and analyzing the reflections.

Satellite Any object that orbits another object in space, whether natural (such as a moon) or artificial, such as a space station. Often used for artificial or man-made objects orbiting Earth, such as communications satellites, comsats, and weather satellites.

Sol The scientific name for the Sun.

Solar To do with the Sun.

Solar eclipse When the Moon passes between the Sun and the Earth, blocking out the Sun's light.

Solar System The Sun and all the planets, moons and other objects that orbit around it or each other.

Spacecraft Any kind of vehicle or vessel built for travel in space. This name is often used for a crewed vehicle, one carrying astronauts.

Space probe A small crewless spacecraft sent to explore space and send information or data back to Earth.

Space station A relatively large space base, orbiting the Earth, where people can stay for long periods.

Star A relatively large space object that for part of its existence contains nuclear fusion reactions that produce heat and light, making it shine or 'burn'. There are many kinds of stars, like white dwarfs, brown stars and neutron stars.

Sunspot A darker, cooler area on the Sun's surface.

Supergiant star A very big, bright giant star.

Supernova The massive explosion at the end of a supergiant star's life.

Telescope Used for studying space, a device that makes faraway things seem bigger, Optical telescopes detect light rays, while other kinds detect other forms of electromagnetic rays or radiation, such as radio waves, IR (infra-red), UV (ultra-violet) and X-rays.

Transit When one relatively small space object is seen passing across the face of a larger, farther one, such as the transit of Venus, when it passes across the face of the Sun when seen from Earth.

Trojans Two groups of asteroids following the same orbit around the Sun as Jupiter, one group in front of the planet and one behind.

Universe Everything that has ever existed, is existing, and could ever exist, including all of space and all its contents.

Year The amount of time a planet takes to complete one full orbit around its star.

INDEX

The letters a, b, c, d, following the page number indicate the column (lettering from left to right) in which the information may be found on that page.